Mental Calculation
An Art
Apart

Willem Bouman

© 2020 Willem Bouman
First Edition, March 2020
ISBN 9781731343444

Contents

Preface v
1 Numbers 1
   1.1   Grouping 2
   1.2   Decimals 2
2 Basic operations 3
   2.1   Additions 3
   2.2   Subtractions 5
   2.3   Mixed additions and subtractions 6
   2.4   Working with decimal numbers 7
   2.5   Multiplication 12
3 About the cross method 13
   3.1   Multiplication 2 × 2 13
   3.2   Is the answer correct? 15
   3.3   Multiplication 3 × 3 16
   3.4   Multiplication 4 × 4 17
   3.5   Addition of arithmetic fractions 19
   3.6   Subtraction of an arithmetic fraction. 20
   3.7   Multiplication of fractions 21
4 Modulo calculation, introduction 24
5 Fractions, special 25
6 Divisions 29
   6.1   Now then, division 29
   6.2   The "inverse" tables 30
   6.3   Integer divisions 31
   6.4   Divisions with remainders 36
   6.5   Decimal divisions 37
   6.6   A special division 38
   6.7   Dividing by four 38
7 Chinese Remainder Theorem 40
   7.1   Introduction 41
   7.2   Another example 42
   7.3   An old Chinese problem 43
   7.4   A box of cigarettes 44
8 Structure in the squares 46
   8.1   Again: the structure 49
   8.2   Creating squares 51
   8.3   Squares, differences of 52
   8.4   Squares, sums of 2 54
   8.5   Sum of two squares 55
   8.6   Congrua 59
   8.7   Answers to the Pythagorases 62

8.8	More about sums of squares	62
8.9	Sums of three squares	68
8.10	Questions	75
8.11	Solutions	75
8.12	Sums of four squares	76
8.13	The numbers	76
8.14	A small inquiry: 49,999 as the sum of four squares.	80
8.15	Surprise, Surprise!	83
8.16	Another small examination: 50,000 as sum of four squares	83
9	Integer square roots	85
9.1	Decimal square roots	87
9.2	Multiplying by 100	89
9.3	Very big roots	89
9.4	Why 13th roots?	91
10	Factoring	98
10.1	Smaller prime numbers	100
10.2	Some simple tricks (and not more than that)	101
10.3	The greater work	102
10.4	Methods	103
10.5	Tools	103
10.6	Filtering methods	103
10.7	Something about the squares	104
10.8	Examples of sums of squares	107
10.9	Some more examples	114
11	Structure in the cubes	122
11.1	The cubic fives	124
11.2	Extracting integer cubic roots	128
12	Finding prime numbers	132
12.1	Finding two primes, bigger numbers	134
12.2	Finding three primes of nine digit question numbers	136
12.3	When to stop dividing	138
12.4	Time needed	140
12.5	Where to start??	142
12.6	Time needed?	143
12.7	How many numbers of this kind are there?	143
12.8	How to get the question numbers	144
12.9	Analysis	144
12.10	The "Russian Roulette"	145
13	Modulo calculation, introduction	146
13.1	Restrictions	147
13.2	The moduli	147
13.2.5 An integer square root		152
13.3	Answers	168
13.4	Summary	169

14	Powering	170
14.1	The squares	171
14.2	Creating squares	174
14.3	The fives	175
14.4	The Cubes	177
14.5	Powering, Q&A	178
14.6	Structure in the fourth power	179
14.7	Quantitative powering	189
15	Integer fourth roots with mod 101	193
15.1	How many figures are to be calculated?	194
15.2	At work!!	194
15.3	Extracting integer fourth roots from 29 up to 32 digit numbers	196
15.4	Questions; integer fourth roots	198
15.5	The answers	199
15.6	The fives in the fourth power	201
16	Structures in the fifth power	203
16.1	Numbers ending on 1	204
16.2	Numbers ending on 2	205
16.3	Numbers ending on 3	206
16.4	Numbers ending on 4	206
16.5	Numbers ending on 5	207
16.6	Numbers ending on 6	207
16.7	Numbers ending on 7	208
16.8	Numbers ending on 8	208
16.9	Numbers ending on 9	209
16.10	The end of endless…	209
17	Table powering 10 and higher	212
17.1	About the 10th power	214
17.2	About the 11th power	214
17.3	About the 12th power	215
17.4	About the 13th power	215
17.5	About the 14th power	216
17.6	About the 15th power	216
17.7	About the 16th power	216
17.8	About the 17th power	217
17.9	About the 18th power	217
17.10	About the 19th power	218
17.11	About the 20th power	218
18	Diophantine Equations	219
18.1	How to calculate the mod 27 of a number	223
18.2	$a^3 + b^3 + c^3 = 684$	224
18.3	$a^3 + b^3 + c^3 = 2403$	224
18.4	$a^3 + b^3 + c^3 = 8504$	225
18.5	$a^3 + b^3 + c^3 = 140939$	227

	18.6	To think about	228
	18.7	$a^3 + b^3 + c^3 = 359982$	229
	18.8	$a^3 + b^3 + c^3 = 436510$, the stumble stone	231
	18.9	With the help of modulo 11	233
19		Mixed operations	237
20		Currency conversion	240
	20.1	Calculation of the amount	240
	20.2	Conversion rates	241
	20.3	Again: calculation of the amount	241
	20.4	Calculation with "Dutch" money	242
21		Calendar calculation	243
	21.1	The day of the week	243
	21.2	The months	243
	21.3	The years	244
	21.4	The centuries	246
	21.5	The questions	248
	21.6	The answers	248
22		Four consecutive primes	251
	22.1	About the 4 figure primes	251
	22.2	How to do this?	254
	22.3	13 digits	254
	22.4	Two 14 digit numbers	255
	22.5	Two 15 digit numbers	256
	22.6	Finally two 16 digit numbers	257
	22.7	Finally	257
23		"Der rasche Rückwärtsrechner"	260
	23.1	One common final figure	260
	23.2	A bit of algebra	261
	23.3	A bit bigger size	262
	23.4	And now, the real work!	264
	23.5	Finally	265
24		Percent calculations	266
About the author			268
Word of recommendation			270
Thanks to…			271

Preface

You have bought a book with the title "Mental Calculation, an Art Apart". From this title it should be clear that the use of any calculating device is strictly prohibited by the author. The reason is this: a machine indeed always gives an answer but never any insight into how that answer is reached. The number of authors who reject the use of a calculating machine – from as young an age as possible – is increasing rapidly. To put this into context, there are no bad words about the use of pen and paper.

Mental calculation is a skill, and this means that practice is essential if progress is to be made. In this book you will read about theories and principles on which the practice is based, these are aimed at enriching your insight into numbers.

As this book regards mental calculation, you'll find only a minimum of mathematics – and only where it is unavoidable. Unfortunately there is a wide spread misunderstanding that mathematics and mental calculation are interchangeable activities. Believe me, they are NOT. There are lots of mathematicians who are excellent in their job, and yet are not so excellent in making standard calculations in their heads. And of course the reverse is also true!

As your training progresses you will become more and more confident with your own skills and you can thus reduce this effect.

I cannot fail to quote a talk I once had with a man who is known to me. He was very busy with pushing buttons and it was obvious he had seen the astonishment in my eyes. "Yes, mister Bouman, you possibly think this man is getting lazy. That's a mistake. By putting all these numbers in my mobile phone I create room in my memory for other things". I answered that he was allowed to decapitate for free if ever this will happen.

There was a suggestion that I should write more than one book and to divide everything into, for example, three books. I rejected this on account of a saying from my English mental calculator friend George Lane "all the operations are interlinked" which I can only confirm.

With this in mind you should not expect an answer that one operation or another is specifically good for, since everything coheres with everything else. You might see reading this book as walking through an orchard; there are many apples, but you do not eat them all.

In this book you will see a number of calculation methods which are not described in any ordinary schoolbook, so whatever else happens at least you will enrich your knowledge.

In numerous articles and books you can read how the quality of the human brain is being ruined by the use of calculating machines. These include "Digital dementia and how we destroy our brain" written by the German psychiatrist Manfred Spitzer, edited by Atlas contact in 2014 and "Don't let your brain sit down", written by Prof. Dr. E. Scherder, edited in 2014 by Atheneum.

Of course at some moment you'll make an error. To find out if your answer is correct, it is much better to check this with modulo calculation rather than the use of a machine. And the more mistakes you make in the beginning, the fewer you'll make after some time.

Particularly, the first chapters should be worked through first; they contain basic principles and operations. After that you can make your own choice.

As well as the increase of your calculating abilities, your practice automatically increases the capacity of your memory generally: recent examinations have proved clearly that the overall capacity of your brain increases when it is better trained and more active, a positive side effect.

We are overwhelmed with numbers every day. Have a look on automobile test reports, almost everything is measured. Why? To give an objective aspect; figures do not lie. In a football match you have the number of passes, percentage ball possession, the distance a player has run, you name it.

A quote of Prof. Aitken during a lecture at the university of Edinburgh in 1959: "Familiarity with numbers acquired by innate faculty sharpened by assiduous practice does give insight in profounder theorems of algebra and analysis."

Pay attention to the combination of talent and assiduous training!

Surely not all the readers will remember the nines test, as a method to check yourself if a multiplication was made correctly. This is an example of "modulo calculation"; there is a special chapter for this.

What then is calculation? In my feeling it is dealing with the numbers which are poured out over our heads every day. Our life is quantitatively destined: there are 24 hours in a day, of which some we need to sleep. We get a reward for our labour, which is always too few for all our wishes, so we have to make intelligent choices et cetera. Just as a carpenter for his daily work needs a folding rule, we need calculation.

At times one can read in the newspaper that the teachers cannot calculate as well as their pupils. Then there is something fundamentally wrong: If you have to explain for a metre, you have to know for a kilometre. This is "over knowledge"; If this is not there, tuition will be fruitless.

An example from the daily life: our dog had a tartar treatment, the invoice was € 155. It was during the week of the teeth of the dogs, and there were chewing sheets for the prevention of tartar. For a special deal there were two for the price of one; each pack of seven sheets, price € 11.55. Well if I could spare our doggy for this treatment, I bought two packs. Coming home I saw weight 143 grams, and then my built-in calculator started up: this means that a kilogram of this stuff would cost over € 80,-. And a single sheet costs € 1.65, with each day one piece. I asked the butcher what the price was of his most expensive meat; € 40,- for rump steak. "You did better to give your dog rump steak than that stuff Sir." This treatment took place every 18 months. This being 548 days would mean 548 × 1.65 = € 904.20; about as much as six treatments. All my calculations were put in a mail and the veterinarian suffered a serious embarrassment. The result: He returned his complete stock of the stuff.

1 Numbers

Although this book is a book about mental calculation, with inevitable references to mathematics, we'll follow the existing lines and will use the same symbols and abbreviations as they are used in mathematics.

So a rational number is, according to mathematics, the quotient (Latin: ratio) of two integer numbers of which the second is not zero. A collection of rational numbers is generally written as Q.

The rational numbers are a part of the real numbers ('R') and include the integer numbers ('Z'). So every integer number is also a rational number and every rational number is also a real number.

Examples of rational numbers are: 1/2, 2/7, 9/4, -2/5, -2 1/2. Every integer number is rational too, e.g. 1 = 1/1 = 3/3, 14 = 14/1 = 56/4, et cetera. Every decimal number with a finite number of decimal places is also a rational number:

$$0.5 = \frac{1}{2}$$

$$0.17 = \frac{17}{100}$$

Not every rational number can be written as a decimal number with a finite number of decimals: For example:

$$\frac{1}{3} = 0.3333$$

$$\frac{15}{7} = 2.142857\ 142857\ 142857\ ...$$

Each decimal number has an infinite number of decimals, the latter one however with a repeating pattern. This is called a repeating fraction. It can be proved that every rational number in the decimal system "behind the comma" has a finite number of figures or an infinite number of figures, in which a pattern is repeating.

If a number with an infinite number of decimals does not have a repeating pattern, this number is irrational. This is the case with, amongst others, $\sqrt{2}$ and π. This will not be discussed in this book.

1.1 Grouping

We follow the usual grouping of magnitude from units to tens, hundreds and so on. This grouping can be written as powers of ten.

10^0	Units	10^5	Hundred thousands
10^1	Tens	10^6	Millions
10^2	Hundreds	10^7	Ten millions
10^3	Thousands	10^8	Hundred millions
10^4	Ten thousands	10^9	Billions

1.2 Decimals

To make things clear: in this book the part of the number behind the comma is considered as the decimal part. If an answer is asked in two decimal places it can for example be written as 12.34. If three decimal places are required the number can be 12.340, and in four decimal places it can be 12.3400.

2 Basic operations

2.1 Additions

The most simple operation of mental calculation is said to be addition. Nevertheless it should be done with maximum concentration and exactitude: no one makes a profit from wrong results.
Addition is the base of multiplication, which we can describe as an accelerated way of adding.

A notion which is also very important is "**complement**". Now we restrict to the use of it in additions, when e.g. the structures in the numbers will be described, you'll see this notion is indispensable. You should get confident with this notion, it gives a "boost" to your abilities.

The complement of any number – generally by the mathematicians denoted as N – is the number which has to be added to obtain the smallest power of 10 which is higher than N. For example: the complement of 79 is 21, of 321 is 679, of 1982 is 8018.
Did this also attract your attention? The cashier in the supermarket gives you the change by working to the complement of the amount of money you gave her.

Looking from right to left, we see that in the complements the sum of the units is 10, and because of the steady carry over the sum of the remaining digits is 9.
We start with integer numbers, later on come the fractions. The best way to work is by breaking down each number in manageable steps. Generally an addition of two 2 digit numbers is called "a two plus two".

Example 1
$$41 + 53$$

The best way to do is work from right to left, to avoid confusions. In our brains we do 4 1 + 5 3. As 1 + 3 = 4 – the units do not exceed a ten, so there is no "carry over", or 'carry' for short. The last digit of the answer is 4. So we can simply add 4 + 5 = 9, the tens. Again we do not exceed a ten, again there is no carry, so the final answer of 41 + 53 = 94.

Example 2

$$79 + 48$$

Keep in mind that regardless the number of numbers to be added, the maximum of hundreds to carry can never be more than the number of numbers. E.g. the sum of 79 + nine other two digit numbers will always be < 1,000. We work again from right to left and add 9 + 8 = 17. The last digit is 7 and we have to carry 1 to the addition of 7 + 4, the tens. Here again we exceed 10, as 7 + 4 = 11 + 1 carry to make 12 which brings the final answer to 127. Doing things like this we make a combination of addition and subtraction.

Here we can use the notion of the complement; the complement from 79 to 100 is 21, which we subtract from 148 and thus find the answer 127.

Example 3

$$27694 + 1841 + 85419$$

The best thing to do is again breaking down the numbers in groups of 2 digits, from right to left – as always – so that we get:

- 94 + 41 = 135 + 19 = 154, answer so far -----54, carry 1.
- 76 + immediately take the carry 1, so that it will not be forgotten! And now have 77 + 18 = 95, nothing to carry yet,
- 95 + 54 = 149, answer so far ---4954, and 1 to carry,
- 2 + 8 = 10 + 1 carry, so the final answer is 114954.

27694 →		2 + 1	76 + 1	94
1841 →			18	41
85419 →		8	54	19
Result		11	(1)49	(1)54

Example 4

$$7428652 + 319847 + 6561630 + 872349$$

Of course people with more experience here can work in groups of 3 digits at a time.

- 652 + 847 + 630 + 349 = 2478, write 478 + carry 2, answer so far ----478
- 428 + 2 carry + 319 + 561 + 872 = 2182, write 182 and carry 2, answer so far ---182478
- 7 + 2 carry + 6 = 15, so final answer 15182478.

If one does not feel confident with steps that big, it is better to take smaller ones and work in groups of two digits at a time or even one. E.G. bookkeepers of my generation – I am from 1939 – are faster in additions by heart than much younger people with a calculator.

7428652 →	7 + 2	428 + 2	652
319847 →		319	847
6561630 →	6	561	630
872349 →		872	349
Result	15	(2)182	(2)478

2.2 Subtractions

Example 1

$$72 - 34$$

Again we work from right to left, and see immediately that 4 is bigger than 2. The simplest way to work is firstly to add 2 to 72 and get 74. 74 – 34 = 40, and after subtracting 40 – 2 the answer 38 is found.

Example 2

$$1869 - 628$$

Immediately is to be seen that 69 is more than 28, with the result 41. In the same way 18 is more than 6, result 12, final result 1241.

1869 →	18	69
628 →	6	28
Result	12	41

Example 3

$$182634 - 76895$$

- 34 – 95 would give a negative result as 95 is more than 34, so we have to "borrow": (1) 34 – 95 = 39 and 26 is reduced to 25. Answer so far 39.
- 25 – 68 would give a negative result either, so again we borrow: (1)25 – 68 = 57 and 18 gets 17. Answer so far 5739.
- 17 – 7 = 10, so the final answer is 10 57 39.

Of course subtractions too can be broken down in groups of 3 digits.

182634 →	18	26	34
76895 →	7	68	95
Result	10	57	39

Example 4

$$354794236 - 264183579$$

- As $579 > 236$ we have to borrow: $(1)236 - 579 = 657$ and 794 gives 793. Answer so far -------657.
- $793 - 183 = 610$; no borrow. Answer so far ----610657.
- The last operation is $354 - 264$. Here again we have to borrow, as $64 > 54$. Final answer 90610657.

354794236 →	354	794	236
264183579 →	264	183	579
Result	90	610	657

2.3 Mixed additions and subtractions

Here surely you should "look before you leap", as errors can easily be made.

Example 1

$$94 - 36 + 79$$

As 6 is more than 4 we have to borrow from the tens, subtracting $96 - 36 = 60 - 2 = 58$, the first part of the answer. Adding 79 to 58 means we have to carry a ten. Adding 79 can be replaced by adding $80 - 1$, so that we get firstly 138, after subtracting 1 our final answer is 137.

Example 2

$$364 - 88 + 187$$

Subtracting firstly 100 gives 264, then adding 12 gives 276. Then adding 187 can be done by firstly adding 200, giving 476, then subtracting 13 results in 463. This way of working increases the experience with the complements.

Example 3
$$8127 - 874 + 793 - 684$$

Step 1. If we start with looking, we'll see easily that in the last part of the question + 793 – 684 results in 109.

Step 2. Add 8127 + 109 = 8236.

Step 3. Subtract 8236 – 874. Take 900 to subtract and get 8236 – 900 = 7336, and add 26 to finish with 7362.

Example 4
$$8554789 - 4522564 + 7552486 - 5458523$$

Make a subdivision as follows 8 55 47 89 – 4 52 25 64 + 7 55 24 86 – 5 45 85 23.

Let's start to look first: 89 – 64 = 25, 86 – 23 = 63, 25 + 63 = answer so far, ----88. We now have the units and tens.

Now the hundreds and thousands: 47 – 25 = 22 + 24 = 46.

Next 46 – 85 and do 46 – (1)00 = 46, add 15, get 61 and borrow one 'hundred' from 55 to then get 54.

Answer so far ----61 88.
54 – 52 = 2 + and now combine + 55 and – 45, = + 10 (+ 2 we had already) = 12.

Answer so far 12 61 88. There is nothing to carry.

Finally we calculate the last figure which we find – remember, there is nothing to carry – by 8 – 4 = 4; 4 + 7 = 11; and 11 – 5 = 6 and we have the complete answer: 6 12 61 88.

2.4 Working with decimal numbers

In our daily life we are confident with working with decimal numbers, regardless the currency we pay with, is only rarely the price of an article is rounded on an integer amount of currency. The prices are almost always to 2 decimal places, so that adding them is hardly different from adding bigger numbers, be it then that a comma or decimal point has to be considered. So let's start with the addition of our bill in the supermarket which could give these prices.

Here is your shopping bill:

€ 1.42	€ 2.77
€ 0.87	€ 4.29
€ 2.69	€ 8.48
€ 1.98	€ 2.19
€ 2.41	€ 6.42
€ 3.31	€ 4.45
Total	€ 41.28

Adding of these amounts gives the sum of € 41.28. What would be our inaccuracy if we rounded all the amounts according to the rules, which means < .50 to .00 and > .50 up to the next euro?

Well, this gives us a total of € 39.00, so our inaccuracy is ± 5 %, rounded, in the advantage of the shopper.

€ 1.00	€ 3.00
€ 1.00	€ 4.00
€ 3.00	€ 8.00
€ 2.00	€ 2.00
€ 2.00	€ 6.00
€ 3.00	€ 4.00
Total	€ 39.00

To give an impression about the effect of rounding on the supermarket bill a table was made about 331 shopping trips.

SMP means Supermarket Price, RP €1 = rounded price in integer Euros, the difference means difference in the advantage of the supermarket. As the supermarket is leading, a positive difference means that by rounding we had to spend more money, a negative percentage means that by rounding the supermarket received less money. There is however one peak of 8.5%; our overall inaccuracy over 331 shoppings is 0.4%. Anyhow, even in the case of the biggest inaccuracy, it is better to have an inaccurate idea about the bill of your supermarket than no idea at all!

After having made the first table the question rose "if the rounding would be made less roughly, in units of 5 Eurocents instead of integer Euros, would there be a significant difference?"

For most shoppers rounding to the next euro gives a satisfactory estimate.

It seems to be that rounding is mostly in the advantage of the supermarket, however,

Rounding to integer Euros

shoppings	SMP	RP €1	difference
22	€ 42.81	€ 43.00	0.4%
35	€ 48.88	€ 49.00	0.2%
34	€ 82.26	€ 82.00	0.3%
31	€ 55.75	€ 56.00	0.4%
26	€ 62.96	€ 63.00	0.5%
28	€ 45.47	€ 45.00	-1.0%
20	€ 28.72	€ 29.00	1.0%
32	€ 54.51	€ 55.00	0.9%
28	€ 59.46	€ 59.00	-0.8%
28	€ 54.46	€ 54.00	-0.8%
27	€ 71.66	€ 72.00	0.5%
20	€ 42.10	€ 42.00	-0.2%
331	€ 649.04	€ 649.00	
Advantage Supermarket		€ 0.04	0.006%

Rounding to 5 Eurocents

shoppings	SMP	RP €5ct	difference
22	€ 42.81	€ 42.80	0.0%
35	€ 48.88	€ 48.90	0.0%
34	€ 82.26	€ 82.25	0.0%
31	€ 55.75	€ 55.75	0.0%
26	€ 62.96	€ 62.95	0.0%
28	€ 45.47	€ 45.45	-0.0%
20	€ 28.72	€ 28.70	-0.0%
32	€ 54.51	€ 54.50	0.0%
28	€ 59.46	€ 59.45	-0.0%
28	€ 54.46	€ 54.45	-0.0%
27	€ 71.66	€ 71.65	-0.0%
20	€ 42.10	€ 42.10	-0.0%
331	€ 649.04	€ 648.85	
Advantage Customer		€ 0.19	0.03%

the differences are small.

Let's not forget: in the Hague there is a book shop with nine chain stores. The amount for the control of the change of this chain is calculated at € 65,000,- a year.

By the way: the 70+ generation of bookkeepers did not have any calculation machine, so they worked only by hand. It is surprising to hear that they, regardless of their age, can still can add a row of fifty numbers faster than younger people with a machine, even if it concerns bigger amounts, e.g. five or even more digits – faultlessly!!

For the record: in the daily practice the supermarket only rounds one time, after having added the total shopping bill. Then the difference is still lower!

2.4.1 Decimal additions

In fact adding a supermarket bill is a decimal addition, but in this case this definition is used for additions in which the numbers have a different number of decimal places. The leading principle in this type of questions is that the biggest number of decimals prevails.

Example 1

$$312.41 + 581.689 + 649.1$$

In this question the .689 is leading: the other numbers have 2 and 1 decimal places. This means that the .41 of the first number should be altered in .410. We can, without problems, add as many zeroes to a decimal number as necessary, as they do not affect the value of a number. The .1 of the number 649 now becomes .100. Adding the decimal parts we see quickly that there is a carry-over of 1 in the integer part, as 41(0) + 689 gives 1,099, and this increased with,100 does not exceed the following 9. .100 + .689 + .410 results in .199 which is the decimal part of the addition. The non-decimal part: 649 becomes 650 because of the carry-over. Then 650 + 581 = 1231, and without the need of the complement we do 1231 + 312 = 1543, so that the final answer is 1543.199.

2.4.2 Mixed addition and subtraction of decimal numbers

Example 2

$$124.8421 + 36.859 - 37.6259 + 29.7325 - 48.866 + 361.612$$

Before starting with the calculation we think and have an overview of the question. It is logical to start with the most decimal places, i.e. 4. We have 0.8421, 0.6259 and 0.7325

0.8421	
0.6259	−
0.2162	
0.7325	+
0.9487	
0.6050	+
1.5537	

As the differences are ± 1.000 and the maximum is ± 2.000 we start with the subtraction as given. We subtract 6300 and get 2121 and add 41 to get 2162. Then adding 7325 results in 9387. As we stay within four d.p. there is nothing to carry over.

In the three d.p. we have + 0.859 and − 0.866, which results in -0.007 question is 1.5537 so that we now can continue with the non-decimal 124 + 36 − 37 = 123. 29 + 361 = 390 + 123 = 513 + 1 carry-over = 514. 514 − 48 = 466, so the final answer is 466.5537.

Example 3

$$127.893 + 13.4782 - 23.72 + 34.9733 - 16.279 + 136.8721$$

0.4782		This is the four decimal part
0.9733		
1.4515		
0.8721	+	
2.3236		

0.2790		This is the three decimal part + the two decimal part
0.7200	+	
0.9990		
0.8930	-	
0.1060		

2.3236	
0.1060	-
2.2176	-

127		This is the integer part, the addition
13		
34		
136	+	
310		
2	+	
312		

23		This is the integer part, the subtraction
16	+	
39		

312		
39	-	
273		The final answer 273.2176.

2.5 Multiplication

We write September 1972. In my father's shop I can do it very well…………
The management had appointed me to give basic tuition to twelve trainees in the tyre business who aspired after a job as a salesman at Michelin. Not because of my calculation abilities but because of my professional knowledge.
But, if whether you like it or not, calculation is also a necessity in the tyre business. Not only because of the discount, no-one on the Western hemisphere buys a tyre for the list price, but it is also important to be able to calculate the axle load of a vehicle to give a client some well substantiated advice for which tyre to choose. And there is more to mention.
By the way, no one had a calculating machine.
For your reassurance; a certificate in higher maths is not required.

In the group one of the guys, the always friendly Tim, attracted my attention because of his very bad calculation work; in a chat with a tyre dealer he would fail dismally in no time.
"But mister Bouman, my father has a dairy, I help him every Saturday and then I can calculate very well"
"Okay Tim, now a milk question especially for you. Three bottles of milk cost 1.19 each and two packs of custard cost 0.88 each. What do I need to pay to you?"
After long thinking – much too long in my feeling – came the answer "6.93".
"No Tim, you do not get that from me, it is too much"
"Sorry sir, I made a mistake, it should be 6.73"
"Dear Tim, that is too much too"
Tim is thinking.
"No, it should be 6.53"
"No Tim, even that is too much. In the meantime I state cheerlessly that even in your business you made three major calculation errors"
"May I try another time?"
"Yes"
"It is 5.93"
"No Tim, this is also incorrect. $3 \times 1.19 = 3.57$; $2 \times 0.88 = 1.76$, together 5.33 and no cent more"
A voice from the group: "Tim you should do like the cheesemonger: "Madam, may it be a little bit more"
Yes, keep then your eyes dry.....

For completeness: After the theoretical course came a stage with an experienced salesman. It was his task to examine if the trainee was capable for the profession of a representative. Unfortunately, Tim appeared to be "insufficient in all aspects"; he was not allowed to continue the training and had to apply for another job.

3 About the cross method

All the mental calculation prodigies use the cross method for multiplication. This has been bethought by the India people, in about the year 1100. It is remarkable that this way of working never has been introduced in the school teaching.

The name of the person who bethought it, is unknown. It seems that the cross method came into the sleep of the Sleeping Beauty, until Dr. Ferrol in the 20th century described it in his book "Das Ferrolsche Neue Rechenverfahren" (The new Ferrol Calculation Method). This method is based on the principle that in the multiplication of numbers consisting of more figures, the result is obtained by multiplying all the parts of the one number with all the parts of the other number.

3.1 Multiplication 2 × 2

Much has been written about multiplication by means of the cross method in many books many chapters. In the books for school tuition the word "cross method" is never mentioned. As far as I know – and I have spoken with almost all mental calculators of the top level – no other method is used by them.

We start with the working out of a classical multiplication: 23 × 89, in which the 3 and 9 are the units and the 2 and 8 are the tens. We do consecutively:

```
       2 3
       8 9  ×
     ─────
       2 0 7
   1 8 4 0
   ─────
   2 0 4 7
```

- Multiply 9 × 3, 7 units + = 27 units × units carry 2
- Multiply 9 × 2 = units × tens + carry 2; this is 20, and
- You now have 207.
- Write a 0 below the 7, as the 8 represents the tens.
- Multiply 8 × 3 = tens × units, = 24, write 4 + carry 2.
- Multiply 8 × 2 = tens × tens, = 16 + carry 2 is 18.
- Then add 207 + 1840 to get 2047.

If we analyse what we just did, we'll see that in the above mentioned multiplication the following happened:

- The units are mentioned four times
- The tens are mentioned four times

So we may conclude we did our job correctly.

10^1	10^0
b	a
d	c
	ac
bc +	ad
	bd

Applying the cross method in a correct way means that we have to do the same operations, only we do this in a different way.

Here we see the working out of the cross method, the column 10^0 means the units as $10^0 = 1$, the units; 10^1 means the tens. The operations are the same, only the working order is different. $a \times c$ = units × units + carry, bc + ad means units times tens + carry, and bd means tens × tens, + carry. The best thing to do is immediately take the carry, so that this important element will not be overlooked. The big advantage of this way of working is that the results are built up gradually and that eventually can be worked without writing intermediate calculations. In mental calculation tournaments these are strictly forbidden and would mean that the answer given is considered as wrong. Moreover it is of paramount importance that with the cross method the memory is trained intensively.

Following the cross method we do now 52 × 79:

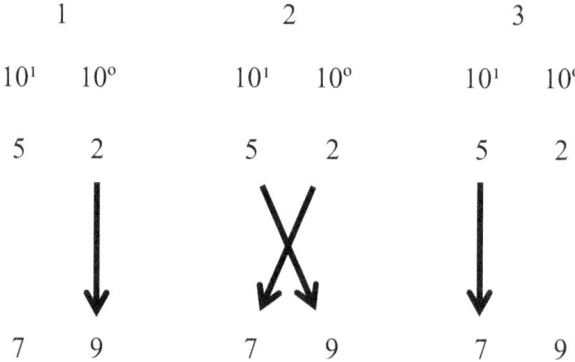

We work out of this multiplication, in which the 10^0 represents the units and the 10^1 represents the tens.

1. The units: 2 × 9 = 18, write 8, carry 1. Answer so far ___8.
2. The tens: 5 × 9 = 45 + 1 carry = 46, 46 + 2 × 7 = 60; write 0, carry 6, answer so far __08.
3. The hundreds: 5 × 7 = 35 + 6 carry = 41. Our work is complete.
4. Answer 52 × 79 = 4108

Have a look at step 2: There is a good reason for multiplying first 5 × 9 and after that adding the carry-over of 1. The aim of this is firstly to eliminate the odd numbers and then continue with even numbers. In practice it has appeared that this is the best way to work.

10^3	10^2	10^1	10^0	
		5	2	
		7	9	
		1	8	2×9
	6	0	8	$1 + 5 \times 9 + 7 \times 2$
4	1	0	8	$6 + 5 \times 7$

The cross method can be extended to any number of figures. If only the principles for the multiplications of polynomials are respected nothing can go wrong. During the Mental Calculation World Championship for multiplication the are 10 questions of multiplication of eight digit numbers. The organiser of this event, Mr. Ralf Laue, told me I am the only one who works with two digits at a time. Why this? I know all the 2 × 2 multiplications by heart since my ninth year, so why not? For greater clarity a 4×4 multiplication is worked out.

Operation	Carry	2389 × 4567
67 × 89 = 5963	59	63
67 × 23 = 1541 + **59 = 1600** 45 × 89 = 4005 + **1600 = 5605**	56	05 63
23 × 45 = 1035 + **56 =**		1091 05 63

Practical hint: always look at the numbers to be multiplied. In this example e.g. 67 × 89 = 5963. If the carry 59 is combined with 67 × 23 = 1541, we have 1600, to which can easily be added 45 × 89. The time I eventually win with the 2×2 multiplications is lost by adding the bigger numbers.

3.2 Is the answer correct?

Please, bethink, we talk about the years ± 1950, shortly after the stone age, there were no machines whatsoever, simply the handwork. How do we check if the answer is correct? We could do the same multiplication another time, with the risk we make the same mistake another time. It is not known which brilliant person has discovered this; we are now talking about the nines test, in fact modulo nine calculation. This means that our results are referred to the number nine, as follows, we take as an example the above mentioned multiplication 52 × 79.

We determine the 'nines remainder' of 52, which is very simple; we add the individual figures, in this case 5 + 2 = 7 and so 52 is 7 modulo 9.

The same with 79: 7 + 9 = 16 and 1 + 6 is 1 + 6 = 7 too. The number of times, i.e. the quotient, is meaningless; we only concern ourselves with the remainder, in this case again 7.

Next: Multiplication of the remainders, so 7 × 7 = 49, and 49 in turn is 4 + 9 = 13 = 1 + 3 = 4. Keep in mind the remainder, 4.

Now we look at our answer: 4108. Also from this number we determine the nines remainder by adding the individual figures: 4 + 1 + 0 + 8 = 13 ÷ 9 = 1 remainder 4.

We now see that the answer of the multiplication of the nines remainders agrees with the nines remainder of the answer, on the grounds of which we may assume the calculated answer is correct.

The nines test is a part of the so called modulo calculation, the calculation with a reference number, more about this later on.

3.3   Multiplication 3 × 3

	10^4	10^3	10^2	10^1	10^0
			4	5	6
			7	8	9
6 × 9					(5)4
9 × 5 + (5) + 6 × 8				(9)8	
9 × 4 + (9) + 7 × 6 + 8 × 5			(12)7		
8 × 4 + (12) + 7 × 5		(7)9			
4 × 7 + (7)	35				
Answer	35	9	7	8	4

The numbers in brackets are the carry-overs.

This is the way to do it:

Units: 9 × 6 = 54, write 4 and carry 5, answer so far ____4.

Tens: 9 × 5 = 45 + 5 carry = 50 + (8 × 6 = 48), sum 98, write 8 carry 9, answer so far ____84.

Hundreds: 9 × 4 = 36 + 9 carry = 45 + (7 × 6 = 42) + (8 × 5 = 40), overall sum 127, write 7 and carry 12, answer so far ___784

Thousands: 8 × 4 = 32 + carry 12 = 44 + (7 × 5 = 35), overall sum 79, write 9 and carry 7, answer so far __9784.

Ten thousands: 4 × 7 = 28 + 7 carry = 35, answer 359784.

Schematically what happened is to be seen in this diagram, from left to right:

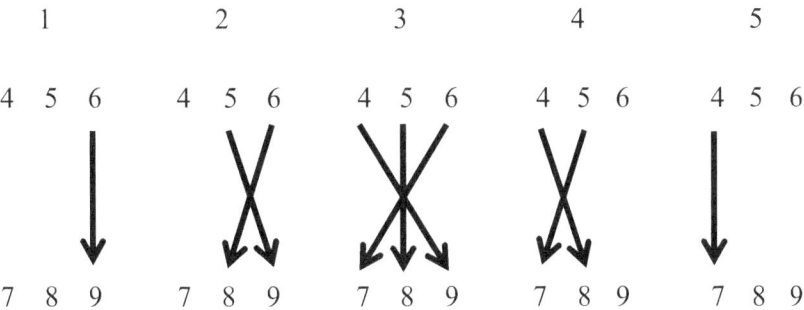

3.4 Multiplication 4 × 4

							10^6	10^5	10^4	10^3	10^2	10^1	10^0
										7	1	8	3
										6	4	9	2
					3 ×	2							6
			8 ×	2 +	9 ×	3						(4)3	
	2 ×	1 + (4) +	9 ×	8 +	4 ×	3					(9)0		
2 ×	7 + (9) ×	1 +	4 ×	8 +	6 ×	3				(8)2			
9 ×	7 + (8) +	4 ×	1 +	6 ×	8				(12)3				
	4 ×	7 + (12) +	6 ×	1				(4)6					
		7 ×	6 +	4	46								
			Answer	46	6	3	2	0	3	6

The numbers in brackets are the carry-overs.

The multiplication of two four digit numbers means only some more effort; the basic principle for the multiplication of two polynomials is maintained.

For clarity, here it is worked out:

Units: 2 × 3 = 6, with no carry, answer so far ____6.

Tens: 2 × 8 = 16 + (3 × 9 = 27), overall sum 43, write 3 and carry 4, answer so far ____36.

Hundreds: 2 × 1 = 2 + 4 carry + 4 × 3 + 9 × 8 = overall sum 90, write 0 and carry 9, answer so far ___036.

Thousands: 2 × 7 = 14 + 9 carry + 6 × 3 + 9 × 1 + 4 × 8 = 82, write 2 and carry 8, answer so far __2036.

Ten thousands: 9 × 7 + 8 carry + 4 × 1 + 6 × 8 = overall sum 123, write 3 and carry 12, answer so far __32036.

Hundred thousands: 4 × 7 + 12 + 6 × 1 = 46, write 6 and carry 4, answer so far __632036.

Millions 6 × 7 = 42 + 4 carry, final answer 46,632,036.

In words, this happens when multiplying:

- units × units give units
- units × tens give tens
- tens × tens AND units × hundreds give hundreds
- units × thousands AND tens × hundreds give thousands
- tens × thousands AND hundreds × hundreds give ten thousands
- hundreds × thousands give hundred thousands
- thousands × thousands give millions

An anecdote

It was during the event for the youth in Münster, November 2017. After the workshop I saw sitting at a table a boy and his mother. As I was interested in which rate the boy had understood the workshop I joined them. The discussion was very lively. I explained to the mother the cross method for 2 × 2 and 3 × 3 multiplications. It seemed to me that the mother understood everything. I asked for some patience while getting my camera. After my return the mother showed me the complete and perfect work out of a 4 × 4 multiplication. "You must be highly educated Madam." "Well I am a doctor and psychotherapist." At the end of the event we said goodbye and I sung for her a German "love" song: "dich werd' ich nie, nie nie vergessen" – "I will never never, never forget you". Her reaction "Neither I'll forget you too".

3.5 Addition of arithmetic fractions

Here also the cross method is an indispensable tool.

$$6\frac{3}{5} + 2\frac{6}{7}$$

Before we are going to calculate we'll have a look at the question: is there some carry? Well, $\frac{3}{5}$ is more than the half and so is $\frac{6}{7}$. We now know that the sum of the two numbers is nine + something.

First step: multiply 5 × 7. We now have the numerator of the fraction 35.

Then, crosswise multiplication of the denominator of the first part of the fraction by the numerator of the other part of the fraction: 3 × 7 = 21, and next the opposite of the other parts of the fractions, and so we have 5 × 6 = 30. Next we add 21 + 30 = 51; the fraction now looks like $\frac{51}{35}$ and we add 2 + 6 = 8 so that we now have $8\frac{51}{35}$. Because with a fraction the denominator always must have a lower value than the nominator, the answer becomes $9\frac{16}{35}$.

A general rule is that a fraction always has to be written in its simplest form. An answer like $6\frac{3}{6}$ will be rejected, this should be $6\frac{1}{2}$.

An example: $6\frac{7}{8} + 4\frac{5}{6}$

Here you can see simply that there will be something to simplify: both numerators are divisible by 2. Both fraction parts are over $\frac{1}{2}$, so there will be a carry.

First step: calculation of the denominator: 8 × 6 = 48.
Second step: crosswise multiplication of successively 7 × 6 = 42 and 5 × 8 = 40.
Third step: 42 + 40 = 82
Fourth step: 82 − 48 = 34, we now have the numerator of the fraction
Fifth step: 6 + 4 = 10 + 1 carry (because of exceeding 48 in the addition of the fraction part) leaves $11\frac{34}{48}$, which will be simplified to $11\frac{17}{24}$.

3.6 Subtraction of an arithmetic fraction.

We use the same numbers, however the question now is

$$6\frac{3}{5} - 2\frac{6}{7}$$

Let's firstly have a look: well, it is as it is but $\frac{3}{5}$ is less than $\frac{6}{7}$. This means that our answer will be 4 minus something, so 3 plus something.

To find the denominator we multiply $5 \times 7 = 35$.

The following steps are equal to these in the addition: $3 \times 7 = 21$, and $5 \times 6 = 30$. We see $21 - 30 = -9$; this means that we will reduce the integer part by 1 (so this will be 3) and we'll use the complement to 35 (which here is 26) so that our answer will be $3\frac{26}{35}$.

Now $6\frac{5}{6} - 2\frac{7}{8}$. Just for a moment we think in the decimal way and see that $\frac{5}{6}$ is less than $\frac{7}{8}$, respectively 0.833 and 0.875. Therefore the integer part of the answer will be 3, + something.

We follow the same procedure as before:

$6 \times 8 = 48$
$5 \times 8 = 40$
$6 \times 7 = 42$

$40 - 42 = -2$, and $-\frac{2}{48}$ gives $\frac{46}{48}$, which we simplify to $\frac{23}{24}$.

In such a case of a negative value we borrow the value 1 from the first number. $6 - 2 - 1 = 3$, so $3\frac{23}{24}$ is the final answer.

3.7 Multiplication of fractions

Here also it is possible to apply the cross method; as you will see this is rather nasty. We take 2 3/5 × 6 6/7, and start with the most nasty part: the fraction part.

$\frac{3}{5} \times \frac{6}{7}$			$\frac{18}{35}$	
$\frac{6}{7} \times 2$	$\frac{12}{7}$	$=$	$\frac{60}{35}$	It is easy to transform 12/7 to 60/35 and then add
			$\frac{78}{35}$ $\frac{78}{35}$	Add
$\frac{3}{5} \times 6$	$\frac{18}{5}$	$=$	$\frac{126}{35}$	It is easy to transform 18/5 to 126/35 and add
			$\frac{204}{35}$	There are no common divisors shared by 204 and 35
		$=$	$5\frac{29}{35}$	Now separate into an integer and a fraction part
2×6			12	The integer part
			$17\frac{29}{35}$	The result

Depending on one's multiplication skills another solution is transforming the 2 3/5 into 13/5 and the 6 6/7 into 48/7. Then multiply the denominators, 13 × 48 = 624; the numerators 7 × 5 gives 35. The result is 624/35 which is simplified to 17 29/35.

Now the question is this: Is there a way to check the answer? Yes, there is, but it is not that simple. It is the nines test.

13 × 48 = 624, 6 + 2 + 4 = 12, and 1 + 2 = 3. The remainder is 3 (mod 9).
5 × 7 = 35 = remainder = 8 (mod 9)
17 × 35 = 595 = remainder = 1 (mod 9)
29 = the remainder = 2 (mod 9)
595 + 29 = 624 = the remainder is 3 (mod 9), so our answer is correct.

The estimation. For this we round the fractions to something more tractable $\frac{3}{5} = \pm\frac{1}{2}$ and $\frac{6}{7} = \pm 1$. We do 2 ½ × 7 = 17 ½; this is an acceptable estimate.

As one cannot walk on only one leg, another example:

$23 \frac{13}{17} \times 19 \frac{8}{11}$

One thing is for sure: the denominator of the answer will be 187.

$$\frac{13}{17} \times \frac{8}{11} \qquad \begin{array}{l}\text{Denominator } 17 \times 11 \\ \text{Numerator } \quad 13 \times 8\end{array} \qquad \frac{104}{187}$$

$$23 \times \frac{8}{11} \qquad \frac{184}{11} \; (\times 17) \qquad \frac{3128}{187} \qquad \text{It is easy to transform 184/11 to 3128/187 and add}$$

$$\frac{3232}{187}$$

$$19 \times \frac{13}{17} \qquad \frac{247}{17} \; (\times 11) \qquad \frac{2717}{187} \qquad \text{It is easy to transform 247/17 to 2717/187 and add}$$

$$\frac{5949}{187}$$

$$31\frac{152}{187} \qquad \text{Now separate into an integer and a fraction part}$$

$$23 \times 19 \qquad \qquad 437$$

$$468\frac{152}{187} \qquad \text{Final answer}$$

The other way: $23 \times 17 + 13 = 404$; so $\frac{404}{17}$. $19 \times 11 = 209 + 8 = 217$, so $\frac{217}{11}$.

The big multiplication 404×217. Be clever, and transform this in $101 \times 868 = 86800 + 868 = 87668$.

The small multiplication: $17 \times 11 = 187$.

We see 876 goes at least 4 times into 187, so the first digit of the answer is 4, remainder 128. Answer so far 4___.

$128 \times 10 = 1280 +$ append the '6' to get 1286.

1286, into which we can fit 6 times 187, the remainder is 164, and the second digit is 6. And we append 8 to get 1648.

Into 1648 we can fit 8 times 187, and the remainder is 152. Final answer $468\frac{152}{187}$.

Again we do the nines test. $404 \times 217 = 87668 = 8 \pmod 9$. As $468 = 0 \pmod 9$ we do not need to make this multiplication, as anything multiplied by 0 is 0, so it's enough to consider only the remainder $152 = 8 \pmod 9$. As the nines remainder of 87668 and the 152 remainder both are 8, we may assume that our answer is correct.

One can also estimate: the answer of the multiplication must be more than $23 \times 19 = 437$ and less than $24 \times 20 = 480$. For this we transform the fractions in rounded ones. $13/17 = \pm ¾$ and $8/11 = \pm ⅔$. The integer part $23 \times 19 = 437$; $+ ¾ \times 19 = \pm 14 = 451 + 23 \times ⅔ = \pm 15$, sum = 466. Although not 100% accurate, this is an acceptable estimate.

About the nines test: the multiplication is 52 × 79, the answer is 4118. 5 + 2 = 7 and from 79 we only take the 7 (to be smart) and multiply 7 × 7 = 49. We ignore the 9 and only take the 4, as we are still smart. The sum of the individual figures of our answer has also to be 4. Now we add: 4 + 1 + 1 + 8 = 14 = 1 + 4 = 5 (mod 9). Here an error has been made and this has to be examined. You know: already earlier this multiplication was made and then the correct answer 4108 was given.

We do the same multiplication and suppose the answer is 4180. We add 4 + 1 + 8 = 13 = 1 + 3 = 4 (mod 9). What we see here is that the nines test suggests a correct answer, while it is incorrect. This can also happen with bigger additions, e.g. ten numbers of ten figures.

4 Modulo calculation, introduction

Later on in this book a whole chapter is devoted to modulo calculation. But because this form of calculation is used much more often, it is practical already now to give some attention to it.
Modulo calculation is a very useful expedient in mental calculation, you will find it many times in this book.

Unconsciously everyone uses modulo calculation every day: if it is 11 o'clock and we have to be somewhere at 14.00 we know we have 3 hours for other activities. Here we calculate using modulo 12.

Another example: it is January 29 and after five days we have to be elsewhere. We then calculate $29 + 5 = 34 - 31$ and find February 3. We have calculated via modulo 31.

What we do in fact is this: we take a certain number as a reference number and we subtract from a result (however many times) that reference number until we have a remainder smaller than that reference number, but not negative. This is the central notion in modulo calculation.

Another example of modulo calculation – surely known by the older generations – is the so-called nines test. This was specially used to check the result of a multiplication. Nota bene: we are here in the era far before the calculation machine. We check the result of the multiplication $59 \times 73 = 4307$. The given answer is 4307. Is this really correct?

We calculate 59 but can ignore the 9 as it is already 0 (mod 9). We only take the 5. Then $7 + 3 = 10 = 1 + 0 = 1$ (mod 9). Next we make the "little" multiplication $5 \times 1 = 5$ (mod 9). We add the individual figures of the answer $4 + 3 + 0 + 7 = 14 = 1 + 4 = 5$ (mod 9). Both of the 'nines remainders' agree so we may suppose that our answer is correct.

If we want to know the mod 9 value of any number we can simply add the individual figures, for example the number 5432. This is $5 \times 1000 + 4 \times 100 + 3 \times 10 + 2 \times 1$. Now is $10 = 1$ (mod 9), $100 = 10 \times 10 = 1 \times 1 = 1$ mod 9; $1000 = 10 \times 10 \times 10$ and by consequence also 1 (mod 9). The mod 9 value of $5432 = 5 + 4 + 3 + 2 = 14 = 1 + 4 = 5$ (mod 9).

5 Fractions, special

At times in the number world we get surprised by two types of very special performances of which we say "How is that possible?".

1. There is a man – almost always – who is able to reproduce a gigantic number of post-decimal digits of the number 'pi'. On October 3 2006 the Japanese Akira Haraguchi succeeded in reproducing 100,000 digits of this number, without any mistake. The speciality of this performance is that in the decimals of pi until there has not – even now – a pattern been found in these digits.
2. There is a man – again almost always – whom is given a fraction e.g. 63 / 97 and he recites in no time the result of this fraction. Only experts in mental calculation recognise something special in this performance; the number of the denominator is practically always the same.

The last performance is not a random surprise, on the base of this performance lies a theory which will be explained here.

Did you ever realise: $1 \div 7$, written as a decimal fraction is 0.142857. There are two things to say about this to remark:

1. If you take another seventh part, you'll see that always the same figures appear: $2 \div 7 = 0.285714$; $3 \div 7 = 0.428571$ et cetera et cetera, all in the same order, the difference is only the place from which the sequence is begun.
2. If you split the decimals in two equal parts, e.g. 142 and 857, and if you add this numbers you get 999

This is a very interesting area, in which only mathematics can help us further. The explanation which follows comes from Mr. B. de Weger, who works as an assistant professor on the technical University of Eindhoven.

We take 7. The secret is in powers of 10 mod 7. Calculate them until you get a result -1 (or 6) or 1 mod 7. Mr Fermat, a mathematician in the 17th century already knew that this will surely happen at a given time. And look:

$10^1 = 10 = 3 \pmod 7$
$10^2 = 100 = 2 \pmod 7$
$10^3 = 1000 = 6 \pmod 7$, and here you are!

This means that $10^3 + 1$ is divisible by 7, the result is 143. And at once we know that 10^6 is the first time that we arrive at a number which is 1 mod 7. And this is linked to the fact that the repeating number 142857 has exactly six figures. What behind this is, this:

$1 / (10^6 - 1) = \frac{1}{999999} = 0.000001000001000001\ldots\ldots$ and so on.

$\frac{1}{7} = 142857 / 999999 = 0.142857142857142857\ldots..$

For any odd and prime number other than 5, at a given moment there is a power of ten which is 1 modulo that number. With this we always find exactly the repeating part of 1/(that number).

Now why we add the two halves to 999999.

The 143 from $(10^3 + 1) / 7$ is 1 more than the 142 of the repeating part, and 857 is exactly $10^3 - 143$. Here is the explanation:

$142857 = 142 * 10^3 + 857$
$= 142 * 10^3 + 10^3 - 143$
$= 143 * 10^3 - 143$
$= 143 * (10^3 - 1)$
$= (10^3 + 1) / 7 * (10^3 - 1)$
$= (10^3 + 1) * (10^3 - 1) / 7$
$= (10^6 - 1) / 7$

Also it is known that $1 \div 13 = 0.076923$. Here we see the same effect: if we split this number into $076 + 923$, the sum of these numbers is 999. The difference is this: if we multiply 0.142857 by 2, 3, 4, 5, or 6, we always get the same figures, albeit in another sequence.

With 0.076923 this multiplying by numbers from 2 up to 12, we get 153846 (dividing this in two equal parts the sum is always 999), then 230769, 307692, 384615, 461538, 538461, 615384, 692307, 769230, 846153 and 923076. You will not be surprised that when adding the "mirror numbers", e.g. $5 \times$ and 8×076923, the sum will be 999999.

Knowing that $73 \times 137 = 10001$ the question came up: if we divide 99 99 99 99 by 73, what do we get then? Well, the answer is 1369863, which can be split up as follows: $1 \div 73 = 0.01369863$, and $1369 + 8630 = 99\ 99$. In the same way $1 \div 137 = 0.00729927$, and $729 + 9270 = $ (again) 99 99.

Table 1

$(10^{16} - 1) \div 17 =$	×	Result
588235294117647	2	1176470588235294
588235294117647	3	1764705882352941
588235294117647	4	2352941176470588
588235294117647	5	2941176470588235
588235294117647	6	3529411764705882
588235294117647	7	4117647058823529
588235294117647	8	4705882352941176
588235294117647	9	5294117647058823
588235294117647	10	5882352941176470
588235294117647	11	6470588235294117
588235294117647	12	7058823529411764
588235294117647	13	7647058823529411
588235294117647	14	8235294117647058
588235294117647	15	8823529411764705
588235294117647	16	9411764705882352

Stimulated by this event we took 1 ÷ 17, which is 0.0588235294117647, which can be split as follows; 05882352 94117647, and what do we see? These two numbers when added result in the total 99 99 99 99.

Multiplying the number 588235294117647 by numbers from 2 up to 16 there are very interesting things to be seen:

We follow the pattern of the individual figures, and what do you see? The pattern of the individual figures repeats, so if you have in your memory the original pattern, it is not very difficult to continue this. In some shows you can see that the "calculator" – generally a memory acrobat – gets presented a fraction as e.g. 23/61. He who knows by heart what is 1/61 can then rather simply "steal the show" in a flash by reciting the complete result.

Behind this is a mathematical theory, which was explained to me by Dr. B. de Weger.

There are two kinds of odd numbers:

1. Those which when modulo calculating in powers of 10 after an even number of times, come on 1, and halfway already were at -1.
2. Those which when modulo calculating in powers of 10 after an odd number of times, come on 1, and never come at -1.

Both give for 1/(that number) a repeating fraction with which the length of the repeating number is equal to the power of 10 which results in 1 when modulo calculating.

Table 2

Prime	N	Prime	N
19	18	61	60
23	22	67	33
29	28	71	35
31	15	73	8
41	5	79	13
43	21	83	41
47	46	89	44
53	13	97	96
59	58		

This is the secret: for every odd number which is not a multiple of five, at a given moment there comes a power of 10 which is 1 modulo that number. With that you always find the repeating part of 1/(that number).

As soon as the length of the repeating number is even, it all goes the same way – always.

To give you an impression about the numbers up to 100 here follows a table which indicates when there is a comparable situation as we saw with 7. For the record, this is the case if the number n is just the prime number minus 1.

For the records: For making this list I used the program 'Mathematica', of which the maker Mr. Steven Wolfram was so kind as to offer me a license for free. Again: thank you very much!!

Although I did not examine all the numbers of the table 2, the result is equal to the calculations from table 1. So, it may be expected that these results are valid for all the numbers from table 1 and all the others which satisfy the requirements as have been mentioned.

Table 3

1/29	0.0 344 827 586 206 896 551 724 137 931
12/29	0.4 137 931 034 482 758 620 689 655 172
17/29	0.5 862 068 965 517 241 379 310 344 827

In table 3 you see the results with 1/29, according to table 2. The figures are for your enlightening separated with spaces, so that adding the groups of three figures is easy – and again you'll see that the order of the figures always remains the same.

To illustrate that at times things go wrong, we take the number 31.

$(10^{15} - 1) \div 31 = $ 032258064516129 032258064516129 032258064516129. Here we see that this repeating number does not have two halves which add to 9999999.

In the same way: $(10^5 - 1) \div 41 = 0.2439$.

6 Divisions

SAMT 1

The courses I gave at Michelin were practically exclusively of technical nature. With one exception, the SAMT, a seminar for the managers in the transport business. Here a big deal of the time was spent on the figures, and you know in the meantime: calculation is an activity I like very much, and over more I can do it very well.

There came a discussion about the item: "regain lost time". And then a question came into my mind, as follows:

There are two distances, each 50 kilometres, from A to B and from B to C.

Given: A car drives from A to B with a speed of 60 km/h and from B to C with 120 km/h.

Question: which is the average speed over the complete distance from A to C, 100 kilometres?

Later, you'll find the solution.

6.1 Now then, division

As dividing is just the opposite operation of multiplication, it lies at hand to assume that the cross method also can be used in the divisions.

"Look before you leap" is also a good principle here. Before starting to calculate it is very useful firstly to have a look at the dividend and the divisor. Generally, the dividend is the bigger number of the two and a good start is to compare these numbers using the same number of figures. Example: 23456789 ÷ 345678. The dividend has 8 figures and the divisor has 6 figures. The first six figures of the dividend are 234567 and make a lower number than 345678, which means that the answer of the division has two figures. In an inverse case, 345678(89) ÷ 234567 we see immediately that six first figures of the dividend make a bigger number than the divisor, so that here the answer will have three figures. For the record: there are no other possibilities!

The magnitudes of the answers are now estimated. In the first case, 23456789 ÷

345678, we can see that the answer is something less than 70, as 7 × 345 = 2415, the answer has 2 figures. If the division is 345678(89) ÷ 234567, the answer has three figures and will be something less than 150, as 15 × 234 = 3510. Here the answer will have 3 figures.

What we have here is the inverse of what happens when we multiply: 12 × 12, two digit numbers results in an answer with 3 digits, 144, and if we take 42 × 42, again two digit numbers, and multiplying results in a four digit answer; 1764.

Besides this one can make an estimation of the answer. In the example it is obvious that the division is not an integer one, since an odd number divided by an even one will automatically result in a remainder.

If we look at 23456789 ÷ 345678, we see immediately that this will not be an integer division: there will be a remainder. And 234 + something divided by 34 + something means that the answer will be 7 minus something.

Of course, we do not start with such big divisions.

In the cross method we work generally with one digit at a time, and we continue by dividing that same way.

SAMT 2

Work out:

A – B	50 km	60 km/h	50 min
B – C	50 km	120 km/h	25 min
A – C	100 km		75 min
Average		80 km/h	

It appears that the average speed seems to be "only" 80 km/h, lower than expected.

Now comes the second question: The speed from A to B remains unchanged, so 60 km/h. The automobile driver however wants to obtain an average speed of 90 km/h over the distance from A to C. The question: At which speed has he to drive the second distance of 50 km to obtain an average speed of 90 km/h over the distance from A to C, 100 km?

Think about this; you'll find the work-out under SAMT 3.

6.2 The "inverse" tables

Many years ago, we learnt the tables from 1 – 10 as the tables of multiplication. In divisions it all goes the inverse way, which we can exploit in a clever way. We know that 7 × 9 = 63, which we can describe a little bit differently as follows: If we multiply a number ending in 7 with a number ending in 9, the result will end in 3. Well then, this is also in force the other way around: if we have a number ending in 3 to be divided by a number ending in 7, the answer will end in 9.

Attention please: In the case of even numbers, it makes sense to look at how many twos there are in each number; for example, 24 ÷ 2. In 24 are 2^3 to be divided by 2^1, the answer contains 2^2 (two factors of 2).

Whoever is dextrous with the multiplication tables of two figure numbers can also exploit this effect. If we know that 27 × 83 = 2241, then also is valid that a number ending in 41 divided by a number ending in 83, will result in an answer ending in 27.

These calculations are used to check an answer, to estimate it, or to exclude eventual mistakes.

A little jump backward: Sometimes you see in the column of the units 10^0, ten to the power zero, in words. This originates as follows: If we divide ten by ten, the answer is 1. If we write this in a mathematic way it is $10^1 \div 10^1 = 10^0 = 1$. If we divide two equal basic numbers we may subtract the exponents. There is something more: based on this reasoning is valid that every number calculated to the power zero equals 1.

6.3 Integer divisions

$$6156 \div 81$$

At school we are taught to work with 615 and then to estimate how many times 81 fits in 615. This is seven times, and we write this '7' immediately. Next is the subtraction:

```
615 -
567
―――
 48
```

Then we append the 6 so that we have 486, and now we have to estimate how many times 81 will fit into 486. Who is clever firstly looks at the 6 and looks how many times 1 fit into 6, and of course this is 6 times. After that you look at 48 and see how many times 8 fits in 48. And this is 6 times too. Now we may conclude that the division 6156 ÷ 81 has the answer 76.

We'll do the divisions like the multiplications and work with one figure at a time.

What we do is split the divisor, in this case in 8 and 1, and only 8 is the divisor. In fact this is not correct, the real divisor is 8+.

This is the way to do it:

- 61 ÷ 8 = 7 r 5. Answer so far 7
- We multiply the remainder × 10, now we have 50, and append the next digit after 61, i.e. 5
- Now we have 55 and subtract 7 × 1 (here we are!) and get 48.
- 48 ÷ 8 = 6 r 0, answer so far 76.
- 10 × 0 = 0 + append the following figure 6, we now have 6
- 6 − 6 × 1 = 0 ÷ 8 = 0, so final answer 76,0

In fact we do not need to do more after 76 because this division is an exact one. if the division is one with a remainder we continue until we have the required number of decimal places.

177157 ÷ 289

Before starting our labour it is useful to have an estimate of the answer. As 177 < 289 our answer will have three digits. If we round 289 to 290 and take the half of this number (145) we can be sure that the answer will be > 5. And adding one more 29 results in 174, the first digit of the answer will be a 6.

Another element: As the larger number ends in 7 and the divisor ends in 9, we can by logical thinking see that the last digit of our answer must be 3. Time for dividing !

Here we take only the 2 as our "main divisor" and subtract the quotients of our partial divisions multiplied by the digits after 2, respectively 6 and 1, because we do the "inverse " cross method.

- 17 ÷ 2 = what? It would not be very clever to write 8 r 1, as we do not divide by 2 only but 2 + a lot, in fact almost 3. The best thing to do is estimate, and if we do 28 × 5 we have 140, which is rather low, and adding another 28 gives 168, which is only a bit lower than 177, so we write 6 r 5, answer so far 6__.
- Remainder 5 × 10 = 50 + append the next digit of the dividend (7) leaves 57 − 6 (first digit of our answer) × 8 (second digit of our divisor) = 9, and 9 ÷ 2 = ??? If we take 4 r 1 our provisional answer is 64. Remainder × 10 = 10 + append next digit of the dividend 1 = 11 − 4 (second digit of our answer) × 8 (second digit of our divisor) − 6 (first digit of our answer) × 9 (third digit of our divisor) = −75, which is surely wrong. Or we do 9 ÷ 2 = 3 r 3 and have provisional answer 63, and then do 31 − 3 × 8 − 6 × 9 = −67, also negative. So we try 9 ÷ 2 = 2 r 5. Provisional answer 62. Remainder × 5 = 50 + append 1 gives 51 − 6 × 8 − 2 × 9 = −66, which is again wrong. Next try: 9 ÷ 2 = 1 r 7. Provisional answer 61. Remainder × 10 = 70 + 1 = 71 − 1 (second digit of

- our answer) × 8 (second digit of our divisor) − 6 (first digit of our answer) × 8 (second digit of our divisor) = 71 − 8 − 54 = 9 ÷ 2 = 3 r 3, the answer so far is 613.
- Remainder × 10 = 30 + append next digit 5 = 35 − 3 (third digit of our answer) × 8 (second digit of our divisor) − 1 (second digit of our answer) × 9 (third digit of our divisor) = 35 − 8 − 24 = 2 ÷ 2 = 0 r 2, answer so far 613,0
- Remainder 2 × 10 = 20 + append next digit = 27 − 0 (of our answer) × 8 (second digit of our divisor) − 3 (last digit of our answer) × 9 (last digit of our divisor) = 27 − 27 = 0, final answer 613.00.

For all clarity: the remainder times 10 + appended number minus a multiplication must always be a positive number!

As generally in competitions it is given if the question is an integer one or one with a remainder we could have stopped after having found 613.

Eventually we can check the answer by using modulo calculation, e.g. mod 9. The dividend is 1 + 7 + 7 + 1 + 5 + 9 = 1 (mod 9), the divisor 2 + 8 + 9 = 1 (mod 9) and the answer 6 + 1 + 3 = 1 (mod 9). We may conclude we have found the right answer.

Doing divisions in this way is rather nasty: it forces the calculator to round down often. In fact one does not divide by 2, but by almost 3. In the following example you will see that if dividing by 28 there is no need for rounding down.

177 157 ÷ 289

Using a divisor with one more digit will bring us a considerable reduction of our efforts. It lies at hand that if we take our divisor as 28, we can start with the dividend taking 177. Over more if we take 28 as divisor, the influence of the 9 in relation to the 28 is considerably smaller.

- 177 ÷ 28 = 6 r 9, answer so far 6
- 9 × 10 = 90 + append next digit of the dividend 1 = 91 − 6 (first digit of the answer) × 9 (third digit of the divisor) = 91 − 54 = 37 ÷ 28 = 1 r 9. Answer so far 61
- Remainder × 10 + appending next digit of the dividend = 95 − 1 (second digit of our answer) × 9 (third digit of our divisor) = 86 ÷ 28 = 3 r 2. Answer so far 613
- Remainder × 10 = 20 + appending next digit of the dividend 7 = 27 − 3 (last digit of our answer) × 9 (third digit of our divisor) 27 − 27 = 0 ÷ 28 = 0 r 0, answer now 613.0

And again: if we know the division is an integer one we can stop after 613. It is useful if the reader studies the working method described above thoroughly, then there will be no disagreeable surprises.

188 315 39 ÷ 6541

It is always good to make an estimate of the answer. As 1883 < 6541 the answer will have 4 digits. As 65 × 3 = 195, the answer will be less than 3, so first digit will be a 2. The last digit of the dividend is a 9, the last digit of the divisor is a 1, so the last digit of the answer will be 9. So we know already our answer to be 2 x x 9.

- 188 ÷ 65 = 2 r 58. Answer so far 2
- 58 × 10 = 580 + append 3 = 583 – 2 × 4 (third digit of the divisor) = 575 ÷ 65 = 8 r 55. Answer so far 28
- 55 × 10 = 550 + append 1 = 551 – 8 × 4 – 2 × 1 (here you see the cross method) = 517 ÷ 65 = 7 r 62. Answer so far 287
- 62 × 10 = 620 + append 5 = 625 – 1 × 8 – 7 × 4 = 589 ÷ 65 = 9 r 4. Final answer 2879
- 4 × 10 = 40 + append 3 = 43 – 9 × 4 – 7 × 1 = 0 ÷ 65 = 0, which can be ignored

The check is by means of modulo calculation: 39 + 15 + 83 + 18 = 155 = 1 (mod 11). 41 + 65 = 106 = 7 (mod 11). 1 (mod 11) ÷ 7 = 8 (mod 11), and as 79 + 28 = 107 = 8 (mod 11) and as 56 (from 8 x 7) (= 1 mod 11) we may conclude that our answer is correct.

391 8019274 ÷ 91 858

Estimation: 39180 < 91858, so the answer has five digits. 391 ÷ almost 92 gives 4+, as 4 × 92 = 368. So 4 is the first digit of our answer. Can we say something about the last digit? 4 ÷ 8 can be either 3 or 8. But 74 is only divisible by 2 (but not by 4) and so is 58, therefore the last digit will be a 3.

- 391 ÷ 91 = 4 r 27, the answer so far is 4
- 27 × 10 = 270 + append 8 = 278 – 4 × 8 = 246 ÷ 91 = 2 r 64, answer so far 42
- 64 × 10 + append next digit 0 = 640 – 2 (second digit of the answer) × 8 (third digit of the divisor) – 4 (first digit of the answer) × 5 (fourth digit of the divisor) = 36 = 604 ÷ 91 = 6 r 58, the answer so far is 426
- Remainder 58 × 10 + append next digit 1 = 581 – 6 (last digit of our answer) × 8 (third digit of the divisor) – 2 (second digit of the answer) × 5 (fourth digit of the divisor) – 4 (first digit of the answer) × 8 (fifth digit of the divisor) = 581 – 48 – 10 – 32 = 491 ÷ 91 = 5 r 36, answer so far 4265
- Remainder 36 × 10 + append next digit 9 = 369 – 5 (fourth digit of the

- answer) × 8 (third digit of the divisor) − 6 (third digit of the answer) × 5 (fourth digit of the divisor) − 2 (second digit of the answer) × 8 (fifth digit of the divisor) = 369 − 40 − 30 − 16 = 283 ÷ 91 = 3 r 10, answer so far 42653
- Remainder 10 × 10 = 100 + append next digit 2 = 102 − 3 × 8 − 5 × 5 − 6 × 8 = 5 ÷ 91 = 0 r 5, answer so far 42653.0
- Remainder 5 × 10 = 50 + 7 = 57 − 0 × 8 − 3 × 5 − 5 × 8 = 2 ÷ 91 = 0 r 2, answer so far 42653.00
- Remainder 2 × 10 = 20 + append next digit 4 = 24 − 0 × 5 − 0 × 8 − 3 × 8 = 0 ÷ 91 = 0 r 0 answer now 42653.000

And again, if it has been mentioned that this is an exact integer division, we can stop after the 3 as we already have reasoned.

Let's check the answer by means of modulo calculation. First check modulo 11. From right to left the dividend: 74 + 92 + 01 + 18 + 39 = 8 + 4 + 1 + 7 + 6 = 26, which is 4 (mod 11). The divisor: 58 + 18 + 9 = 3 + 7 + 9 = 8 (mod 11). We look for a number 4 (mod 11) to be divided by 8 (mod 11). The shortest way: 4 ÷ 8 = ½ + (½ × 11) = 6 (mod 11). Our answer 53 + 26 + 4 = 9 + 4 + 4 = 6 (mod 11), so this is correct.

A bit longer is this way: We have 4 (mod 11) and have to add so many elevens till we have a number which is divisible by 8. After 4 × 11 we have 48 ÷ 8 = 6.

745 28464284 ÷ 84 1293

Again an integer division.

First action: Thinking. 745284 < 841293, so the answer has 5 digits. Estimation: 8 × 84 = 672 and 9 × 84 = 756, so the first digit of the answer is 8. Last digit: 4 ÷ 3 = 8, uniquely. We can even say more about the last digits: 4284 is divisible by 4, so the last digits of the answer are 88 + an odd number of hundreds. As the method is quite the same as described above, the steps will not be explained in such detail.

- 745 ÷ 84 = 8 r 73, answer so far 8
- 73 × 10 + 2 = 732 − 8 × 1 = 724 ÷ 84 = 8 r 52, answer so far 88
- 52 × 10 = 520 + append 8 = 528 − 8 × 1 − 8 × 2 = 504 ÷ 84; here we round down because of the small remainder if we were to take 6 × 84. So we take 5 × 84 r 84, answer so far 885
- 84 × 10 + append 4 = 844 − 5 × 1 − 8 × 2 − 8 × 9 = 751 ÷ 84 = 8 r 79, answer so far 8858
- 79 × 10 + append 6 = 796 − 8 × 1 − 5 × 2 − 8 × 9 − 8 × 3 = 682 ÷ 84 = 8 r 10, answer so far 88588
- 10 × 10 + append 4 = 104 − 8 × 1 − 8 × 2 − 5 × 9 − 8 × 3 = 11 ÷ 84 = 0 r 11, answer so far 88588.0

- 11×10 + append 2 = 112 − 0 × 1 − 8 × 2 − 8 × 9 − 5 × 3 = 9 ÷ 84 = 0 r 9, answer so far 88588.00
- $9 \times 10 = 90$ + append 8 = 98 − 8 × 9 − 8 × 3 = 2 ÷ 84 = 0 r 2, answer so far 88588.000
- $2 \times 10 = 20$ + append 4 = 24 − 0 × 1 − 0 ×2 − 0 × 9 − 8 × 3 = 0, final answer 88588.0000.

In fact we could have stopped after 88588, as the question is presented as an integer division.

Finally the elevens test, from right to left: 84 + 42 + 46 + 28 + 45 + 7 = 7 + 9 + 2 + 6 + 1 + 7 = 32 = 10 (mod 11), 93 + 12 + 84 = 5 + 1 + 7 = 2 (mod 11). 10 ÷ 2 = 5 and our answer is 88 5 88, we can ignore two times 88, as they are 0 (mod 11) the remaining 5 = 5 (mod 11) so we may assume that our answer is correct. As both dividend and divisor are multiples of 9, the nines test is not used.

SAMT 3

Work out of the question "How fast have the second distance of 50 km be driven to increase the average speed over the distance from A to C to 90 km/h?"

A – C 100 km	90 km/h	10/9 hr	66 2/3 minutes
A – B 50 km	60 km/h		50 minutes
B – C 50 km	??? km/h		16 2/3 minutes
B – C 50 km	3 km /minute		**180 km/h**

Our conclusion can only be that lost time cannot be regained. It was funny that later on some managers phoned me back: Driving more quietly does not only save many litres of fuel, but it also works out positively on brake wear and tyres, all of them parts of the variable costs per kilometre.

6.4 Divisions with remainders

$$4273 \div 64$$

- $42 \div 6$ = in fact 7, but as our real divisor is 6+, we round down and write 6 r 6. Answer so far 6.
- Remainder × 10 gives 60 + append next digit after 42 (which is 7) and now we have 67
- $67 - 6 \times 4 = 43 \div 6$ = actually 7 r 1, but as the real divisor is 64 we round

down to 6 r 7. Answer so far 66
- Remainder × 10 = 70 + append next digit after the 7 (i.e. 3) and we get 73.
- 73 – 6 × 4 = 49, and as 49 is smaller than the divisor our final answer is 66 remainder 49.

$$286739 \div 288$$

The answer will have 3 digits as 286 < 288, and it will be close to 1000, for 288000 – 2880 = 285120. It will even be close to 995. Consider that ½ × 288 = 144 + 28512 (= 99 x 288) = 28656.

- 287 ÷ 28 = 9 r 35, answer so far 9. This division is rounded down, as we in fact divide by 28 + a lot
- 34 × 10 + append next digit 7 = 347 – 9 × 8 = 275 ÷ 28 = 9 r 23. Answer so far 99
- 23 × 10 = 230 + append 3 = 233 – 9 × 8 = 161 ÷ 28 = 5 r 21. Answer so far 995
- 219 – 5 × 8 = 179, which is the remainder.
- Final answer 995 r 179

6.5 Decimal divisions

We can also transfer the above mentioned division in a decimal division

$$286739 \div 288$$

- 286 ÷ 28 = 9 r 34, answer so far 9
- 34 × 10 + append next digit 7 = 347 – 9 × 8 = 275 ÷ 28 = 9 r 23. Answer so far 99
- 23 × 10 = 230 + append 3 = 233 – 9 × 8 = 161 ÷ 28 = 5 r 21. Answer so far 995. As you see: until now there is no difference. But be quiet please: this will come soon!!
- 219 – 5 × 8 = 179 ÷ 28 = 6 r 11, answer so far 995.6
- 11 × 10 = 110 – 6 × 8 = 62 ÷ 28 = 2 r 6. Answer so far 995.62
- 6 × 10 = 60 – 2 × 8 = 44 ÷ 28 = 1 r 16. Answer so far 995.621
- 160 – 1 × 8 = 152 ÷ 28 = 5 r 12, answer so far 995.6215
- 12 × 10 = 120 – 5 × 8 = 80 ÷ 28 = 2 r 24, answer so far 995.62152
- 24 × 10 = 240 – 2 × 8 = 224 ÷ 28 = 7 r 28, answer so far 995.621527

Here we can stop. If five decimal places are required, the answer will be 995.62153 because of the rounding of the 7.

6.6 A special division

$$35467 \div 367$$

This division is presented to sharpen up our minds and develop our feeling of quantities and qualities. The central question is this one: Is this an integer division, yes or no, and how can we find our answer without really having to calculate, only by reasoning?

Main question: is this an integer division, yes or no? And why, or why not?

These are our questions:

- How many figures has the answer? Two, as 354 < 367. This is a quantitative answer
- Can we say something about the nature of the figures in the answer? Yes, as we divide xxx67 by a number yyy67 the last figures of the answer should be 01.
- Is then the answer 101? No, as 101 × 367 is without much ado more than 36700 and therefore more than 35467.
- Here we may conclude this is not an integer division.
- What else can we say?
- Well, we apply modulo calculation. 35467 = 7 (mod 9) and 367 = 7 (mod 9). Dividing 7 (mod 9) by 7 (mod 9) will result in an answer 1 (mod 9). If we had 101 as an answer, then when multiplying we would get a number 367, which is 7 (mod 9) and the following number ending in 01 is 901, which, multiplied by 367 will be a shot far over the mark. Multiplying by 101 does not fit, as 101 = 2 (mod 9) and multiplying this with 7 (mod 9) results in a number 5 (mod 9) which conflicts with our initial number.

Anyhow: this division is not an integer one.

6.7 Dividing by four

This real life story happened in a bank, a "money shop".
To optimise the service for the clients there were only a few people in service, so that the waiting times were considerable. A girl of about 16 years was rather stressed: for the next day she had a lot of homework to do.

"Is it really so much? May I ask what kind of work?"
"English, mathematics and economy."
"Wow, that's quite a lot! May I ask which kind of questions you have for

mathematics?"

"Well, for example how much is ten euro's plus twenty five percent. But for this kind of questions we may use a calculation machine."

"Well well, this is really tough!! But could you solve a problem like this by means of mental calculation?"

"Yes, I can. In fact it is rather simple. A quarter of ten is two and a half, so the answer is ten euro and twenty five cents. (€ 10.25)!!"

7 Chinese Remainder Theorem

The Chinese remainder theorem was firstly described in the 4th century BC by the Chinese mathematician Sunzi. It concerns numbers which have no common divider. The theorem gives an answer on questions like the following:

A given number X has the following properties:

- X divided by 3 gives a remainder 2, written as X = 2 (mod 3)
- X divided by 4 gives a remainder 3, written as X = 3 (mod 4)
- X divided by 7 gives a remainder 5, written as X = 5 (mod 7)

Find X.

On the Internet is to be found the mathematic answer on this question, amongst others, a Euclidean algorithm (??????????????????) and simultaneous congruencies (??????????????????); these are far beyond my knowledge.

The solution for this question I found myself in an arithmetic way. This is my way of doing it, completely with the help of modulo calculation and without any mathematic algorithm. Example:

X = 2 (mod 3) and 3 (mod 4). Then X is rather simple to find; we multiply 3 × 4 = 12 and subtract 1 to get 11. We now know that amongst others X = 11 (mod 12), but we are not yet where we have to be.

X is also 5 (mod 7), which is given.

This is the reasoning:

- With 5 (mod 7), we can make the following row: 5, 12, 19, 26, et cetera. If it is like that, we do not need to add more than 12 × 7, for certainly it will repeat. We always come lower than 3 × 4 × 7 = 84, as this is the smallest common multiple of 3, 4 and 7. Starting at 5 we do not need to add more than 11 × 7.
- We also know that we have to add an even multiple of 7 as we start with 5 (mod 12) and we have to obtain 11 (mod 12).
- Starting at 5 we continue in steps of 7 so many times until we have a number which is 11 (mod 12). So we have to gain 6 (mod 12). Now the question is

how many steps of 7 do we need to get to 6 (mod 12)? Well, $2 \times 7 = 2$ (mod 12) and if we take 3×2 we have the 6 we looked for, as $6 \times 7 = 42 = 6$ (mod 12).

We now arrive at $5 + 6 \times 7 = 47$ and all the numbers which satisfy the mentioned requirements are equal to 47 (mod 84).

7.1 Introduction

From 8–10 May 2009 in my house was the annual calculation prodigy weekend. My guests were Jan van Koningsveld, having a lot of world records, and Dr. Andy Robertshaw, a promoted mathematician and mental calculation prodigy and Dr.Dr. Gert Mittring, manifold World Champion.

To discuss was a question as to be found in a tournament; a big division with a remainder. It is to be realised that the only thing which is allowed is the final answer to be written in a box, not a single kind of "help-notice" is allowed.

This was the question: find the remainder of

$$139875219643 \div 121068$$

We factorised: $121068 = 4 \times 27 \times 19 \times 59$. A short explanation:

121068; with the finish of 68. With an even count of hundreds there is no more than 2^2.

If there is a six figure number and the two halves are added, in this case $121 + 068 = 189$, and the sum is divisible by 27, then the whole number is too.

Andy Robertshaw started with the explanation of the Chinese remainder theorem, which indeed is the clue to the solution. He calculated that the dividend = 3 (mod 4), 13 (mod 27), 8 (mod 19) and 37 (mod 59).

During Andy's explanation I got the idea of modulo calculation and worked this out as follows, in the same way as I did above.

Step 1.
$X = 3$ (mod 4) and is also 3 (mod 27). We have to arrive at 13 (mod 27). The question is how to get 10 (mod 27) in steps of 4. Answer: 16 steps, as $16 \times 4 = 64 = 10$ (mod 27), so 16 steps, now take 3 and add 16×4 and we get 67, which indeed = 3 (mod 4) and 13 (mod 27).

Step 2.
The number we are looking for is not only 3 (mod 4) and 13 (mod 27), besides this it is also 8 (mod 19). 67 = 10 (mod 19) and we have to arrive at 8 (mod 19) in steps of 4 × 27 = 108 which is 13 (mod 19). From 10 (mod 19) to 8 (mod 19) = 17 (mod 19), as 10 + 17 = 27. As 3 × 13 = 1 (mod 19), we need 17 × 3 = 51 = 13 (mod 19) × 108 to find the number required. 67 + 13 × 108 = 1471, being indeed 3 (mod 4), 13 (mod 27) and 8 (mod 19). Expressed correctly: 1471 is 1471 (mod 2052), as 4 × 27 × 19 = 2052.

Step 3.
Starting with 1471 we have to find a number which is not only 3 (mod 4), 13 (mod 27) and 8 (mod 19), but moreover 37 (mod 59), in steps of 4 × 27 × 19 = 2052 = 46 (mod 59). As 1471 = 55 (mod 59) we have to gain 41 (mod 59) in steps of 46. 9 × 46 = 414 = 1 (mod 59). As 41 × 9 = 369 = 15 (mod 59) we need to add 15 × 2052 = 30780 + 1471 = **32251,** a number which meets all the requirements; 3 (mod 4), 13 (mod 27), 8 (mod 19) and 37 (mod 59). So the remainder of the division is 32251.

7.2 Another example

Given: 1, 5, 23, 53, 1523, 29243. Question: What is the following number?

Solution: 1 = 1 (mod 2); 5 = 2 (mod 3); 23 = 3 (mod 5); 53 = 4 (mod 7); 1523 = 5 (mod 11); 29243 = 6 (mod 13). So we have to look after a number which, besides the already mentioned properties, is also 7 (mod 17).

Step 1.
1 = 1 (mod 3). We have to go from 1 (mod 3) to 2 (mod 3) in steps of 2. This is 2 steps and we have now 5, which is 1 (mod 2) and 2 (mod 3).

Step 2.
5 = 0 (mod 5) and we have to go from 0 (mod 5) to 3 (mod 5) in steps of 2 × 3 = 6. This will take 3 steps, and we have then 23 which is 1 (mod 2), 2 (mod 3) and 3 (mod 5).

Step 3.
23 = 2 (mod 7) and we have to go to 4 (mod 7) in steps of 2 × 3 × 5 = 30, which is 2 (mod 7). We need one step and then we have 53 which is 1 (mod 2), 2 (mod 3) and 3 (mod 5) and 4 (mod 7).

Step 4.
53 = 4 (mod 7) and 9 (mod 11). We have to go to a number which is 5 (mod 11) in steps of 2 × 3 × 5 × 7 = 1470 = 7 (mod 11). 9 (mod 11) + 7 (mod 11) = 5 (mod 11), just what we need, and 53 + 1470 = 1523.

Step 5.
1523 = 5 (mod 11) and 2 (mod 13) and we have to go to 6 (mod 13) in steps of 2 × 3 × 5 × 7 × 11 = 2310. 2310 = 9 (mod 13). So how many steps of 9 (mod 13) do we need to get to 4 (mod 13)? This will be 12 steps and then we have 1523 + (12 × 2310 =) 27720 = 29243 = 6 (mod 13).

Step 6.
From 29243 = 3 (mod 17) we have to arrive at 7 (mod 17) in steps of 2 × 3 × 5 × 7 × 11 × 13 = 30030 = 8 (mod 17). For reaching 4 (mod 17) in steps of 8 we need nine steps, as 9 × 8 = 72 = 4 (mod 17). Now we take 29243 and add 9 × 30030 = 270270 and then we have the solution; the number wanted is 299513.

299513 can only be written as the sum of $277^2 + 472^2$, so it's a prime number.

7.3 An old Chinese problem

In a syllabus I found the following problem, which was presented as an old Chinese problem.

Three Chinese famers divide the rice they have grown together. They go to markets where different weights are handled. The first farmer goes to a market where they handle units of 83 pounds and has a remainder of 32 pounds. The second famer goes to a market where they work with units of 110 pounds, where he has a remainder of 70 pounds. The third famer goes to a market where 135 pounds is the standard, and he has a remainder of 30 pounds. How much rice did they grow together?

We firstly look after a number which meets the following conditions:

- Divided by 83 there should be a remainder 32
- Divided by 110 there should be a remainder 70

According to "modulo thinking" we start with 32 and want to finish with a number which is 70 mod 110 in steps of 83.

Thinking about mod 110 we start with 32 and must "gain" 38 to finish with 70, so there are 38 mod 110 to gain, again in steps of 83. This means that the number of times has to be a ?6 number, as 6 × 3 ends in an 8. Surprise, surprise!! 83 and 38 are each other's antipodes: 83 + 38 = 121, which is 0 mod 11. This means that the number of times 83 must be 10 mod 11 to result in 38. So the number we need is:

- Ending in 6
- 10 mod 11
- Lower than 83

There is only one number which meets these requirements: 76. We take 32 + 76 × 83

= 6340 and here we have the number which also is 70 mod 110, as 70 + 57 × 110 = 6340.

The second step is to start from 6340 and go to a number which divided by 135 results in a remainder of 30, and to do this in steps of 83 × 110 = 9130.

6340 = 130 mod 135 as 46 × 135 = 6210 and 6210 + 130 = 6340. And 9130 = 85 mod 135 as 67 × 135 = 9045 and 9045 + 85 = 9130.

Now the question is: How many steps do we need to take for starting with 130 mod 135 to finish with 30 mod 135 in steps of 85.

Thinking boldly in mod 135 we now get: 130 + 85 = 215 = 80 mod 135 and 80 + 85 = 165 = 30 mod 135. So now we have what we are looking for: starting with 6340 we take 2 × 9130 and get 24600, and here we are.

We check this and find:

24600 ÷ 83 = 296 remainder 32
24600 ÷ 110 = 23 remainder 70
24600 ÷ 135 = 182 remainder 30

We may now suppose that the famers have grown 3 × 24600 = 73800 pounds of rice.

7.4 A box of cigarettes

During the school holidays one could earn some money by working, for example, in the local chocolate factory. The following happened during lunch time. The regular workers had read about my calculation abilities in a local newspaper. One of these workers presented me a question, which no-one before could solve, and if I succeeded he would give me a box of cigarettes.

This was the question given a number which

- Divided by two gives remainder one
- Divided by three gives remainder two
- Divided by four gives remainder three
- Divided by five gives remainder four
- Divided by six gives remainder five
- Divided by seven gives remainder six
- Divided by eight gives remainder seven
- Divided by nine gives remainder eight
- Divided by ten gives remainder nine

Which is this number? It took some time, some minutes but then the answer came: "the number is 2519".

Then the man got mean: "You must have known the question, no one can solve it."

Of course I denied "No, it is new for me and I want my reward."

His mates got rather angry and pressed him to give me the cigarettes. He gave the smallest one he could; a box of ten.

8   Structure in the squares

Table 1

00	01	04	09	16
21	24	25	29	36
41	44	49	56	61
64	69	76	81	84
89	96			

Table 1 gives you a survey of the possible two final figures of the squares 0–100. For all clarity: if such a square has only an even hundred (EH) the box is light gray, for the squares which have exclusively odd hundreds (OH) the box is made dark gray, and if both is possible the box is blank.

What is so striking in this table? The blank boxes represent the numbers 0 (mod 4), so they are the squares of even numbers. But there is more; we take 04, the square of 2. If we take the following square ending on 04, 2304, the square of 48, we see an odd hundred. 16 is the square of 4, the following square ending in 16 is 2116, the square of 46.

Thinking in terms of 0 mod 16 we'll see that all the squares ending on 16, 24, 56, 64 and 96 will have even hundreds, the squares 0 mod 16 and ending on 04, 36, 44, 76 and 84 will have odd hundreds.

For the other squares this is valid: once the hundred being even, they remain even; once the hundreds being odd, they remain odd. E.g. if we take 09 as the square of 3, 103^2 is $3^2 + (2 \times 3 \times 100) + 100^2$, which we know as $(a + b)^2 = a^2 + 2ab + b^2$. The secret is in the 2ab, which will always results in an even hundred to be added..

To quote the well-known American author Ron Doerfler: "In things I've written, I've pointed out that knowing squares is the single best thing for mental calculation generally, and you must be the world expert in that!".

Thank you, Ron!

At times on the internet we can read a message where a method is asked for – how to learn the squares by heart.
My answer, as always: do this only if you intend to make an intense use of it. And if this is not the case, use your memory for more useful things.
There is in the meantime a well-known saying "If you don't use it, you lose it".

From where my interest in the squares?

We write second class, secondary school, 1952. The tuition in square roots. Our math teacher had a very nice nickname "Zacharias Carpenter", of which I do not remember the origin. He knew all the squares up to 25 by heart and he was very proud of it. Well, why not? In an impulse I started with 25^2, 625 and then $26^2 = 676$. Hey, what's that?? $24^2 = 576$, so the squares of 25 – 1 and 25 + 1 both end in 76. And the squares of 23 and 27?? Surprise, surprise, they both end in 29. And from that moment my interest in the structure of the squares was fired up. Later on my teacher explained to me that this all coheres with $(a+b)^2$ and $(a-b)^2$.

In the second class of my secondary school we started with integer square roots. Already knowing the structure of the squares up to 100 I was anxious to know if such a structure could be found in the squares up to 1,000. There was no table in which they were mentioned, and I resorted to a particular means; writing them out from 1 up to 1,000. To speed up this activity a bit, the less interesting lessons were also used for this goal. Despite my poor mathematical abilities I found a simple way for this work: by adding. $2^2 = 1^2 + 1 + 2 = 4$. Of course in the same way $343^2 = (342 + 1)^2 = 116964 + 342 + 342 + 1$ so 117649.

In fact this was also an application of $(a + b)^2 = a^2 + 2ab + b^2$: the 342 + 342 represents the 2ab.

Owing to my excellent memory – a gift of my Creator – and intensive training, all the numbers were well saved. And after that a start could be made with an inquiry into the structure in them.

Table 2

An aside remark: As I was heavily committed in this matter, and therefore I believe that the squares were saved so well in my memory, it was fully my own work. Looking around me I see heaps of – generally young – people who use a calculator, having no idea about quantities, and by consequence have no "figure fixedness" at all. If later on they see a number which they had seen sometime before, there is no sign of recognition. A senior accountant told me: Younger accountants also do not have "figure fixedness", as they make all their calculations on a machine. Not a single number is put in their memory, they just typed it in the machine.

Basic Number	Square
1	1
2	4
3	9
4	16
5	25
6	36
7	49
8	64
9	81

Table 2 gives us the first and basic insight into what happens when we start squaring.

- In the last figures of the squares we miss the figures 2, 3, 7, 8
- We see that 1 remains 1, 5 remains 5, 6 remains 6
- We see 4 pairs of numbers with the same final digits: 1 and 81, 4 and 64, 9 and 49, 16 and 36

- A "mirror" effect: 1 + 9 = 10, 2 + 8 = 10, 3 + 7 = 10, 4 + 6 = 10

This leads to the question: are there more symmetries in the squares? Yes, there are; see table 3. You'll be surprised too: quite a lot!

Table 3

2^2	**4**	3^2	**9**	7^2	**49**	8^2	**64**
48^2	23**04**	47^2	22**09**	43^2	18**49**	42^2	17**64**
52^2	27**04**	53^2	28**09**	57^2	32**49**	58^2	33**64**
98^2	96**04**	97^2	94**09**	93^2	86**49**	92^2	84**64**

In the vertical columns you see in bold the squares of numbers with the same last digits. In the columns left of them are the basic numbers. Immediately it is seen that the sums of the first and the last number is always 100. This is also the case with the second and third number. Numbers 1 and 3, likewise 2 and 4, differ by 50. Explanation: $(100 - a^2) = 10000 - 2 \times 100 + a^2$, and this modulo 100: $(100 - a^2) = a^2$ mod 100, so the last two figures are equal. If we look at mod 1000 we see even that the hundreds differ an even number, so either both are even or both are odd. We can also have a look at 2 and 52 = 50 + 2, or more generally after a^2 and $(50 + a)^2$. The last one is $2500 + 100a + a^2$. modulo 100; this shows $(50 + a)^2 = a^2$ (mod 100), so a number + 50 being squared has always the same two final figures. Therefore the two final figures always appear in groups of four.

An anecdote

Our math teacher on the secondary school, he has already been mentioned, wrote for his tuition on the blackboard √ 64 304 361. Immediately seeing the answer I shouted through the class room the answer: 8019. "OK, you young man, when you know already everything so well, you are asked to leave the class room".

Also for the other maths teachers at school it took some time before they accepted my special abilities. To their great astonishment I finished my school in the economy-language department and not in the mathematic direction.

It is still thought that that a real calculator is also a good mathematician, and the other way around that every mathematician is a good calculator. Here follows a quote from a booklet, written in 1913 by Dr. Philipp Maenchen: "If one wants a calculation being made correctly, one should not charge a mathematician with it". And "there are for the science meaningful mathematicians who however do not excel in arithmetic calculations".

8.1 Again: the structure

Now we have seen that generally there are symmetries, we want to see more. Table 4 shows us the symmetry of the squares up to 1,000.

Here too the symmetry is visible: when adding the first and last numbers, the second and seventh et cetera you'll see that the sum is always 1,000.

Aside from this I remark that from three digits onward there are still more of this kind of groups to be composed. The five digit numbers not ending in 0 or 5 can be divided into groups of eight numbers which have five identical final digits when squared.

Table 4

11^2	121	23^2	529
239^2	57121	227^2	51529
261^2	68121	273^2	74529
489^2	239121	477^2	227529
511^2	261121	523^2	273529
739^2	546121	727^2	528529
761^2	579121	773^2	597529
989^2	978121	977^2	954529

Let's have a look at the left column, the one of the 121s. We see here the even thousands (ET), in the squares of 11, 261, 739 and 989, and the odd thousands (OT), the squares of 239, 489, 511 and 761. To understand this, we change to mod 16 calculation. The squares of 3, 5, 11 and 13 (mod 16) are all 9 (mod 16), and 11, 261, 739 and 989 are resp. 11, 5, 3 and 13 (mod 16) and ET are 0 (mod 16) and 121 in itself is 9 (mod 16). 239, 489, 511 and 761 are resp. 15, 9, 15 and 9 (mod 16), OT are all 8 (mod 16) + 121 9 (mod 16) result in 1 (mod 16).

The same reasoning is valid for numbers ending in 529. ET 529 = 1 (mod 16). The basic numbers 23, 273, 727 and 977 are resp. 7, 1, 7 and 1 (mod 16), their squares by consequence 1 (mod 16), ET are 0 (mod 16) and here you are. 227, 477, 523, and 773 are resp. 3, 13, 11 and 5 (mod 16), their squares are all 9 (mod 16). 529, 1 (mod 16) + an OT will result in a number which is 9 (mod 16).

Another element which should not be forgotten is this: All the squares of even numbers are of course 0 mod 4. All the odd squares are 1 mod 4. More about this in the chapter "Sums of Squares".

Table 5

14^2	196	514^2	264196
236^2	55696	736^2	541696
264^2	69696	764^2	583696
486^2	236196	986^2	972196

We see in table 4 that up to 1,000 there are 8 squares with 3 equal final digits and again the first basic number + the last one together are 1,000, 2nd + 7th is the same et cetera et cetera.

An exception should be made for the even numbers, which can differ 250. And also e.g. $62^2 = 3,844$ and $312^2 = 97,344$. Here the hundreds differ by 5 (500).

These differences depend on the basic numbers. When the basic number is only divisible by 2, by consequence the square will be divisible by only 4. In fact we should now calculate mod 16. (H)96 is only divisible by 8 with an even hundred, 196 is not divisible by 8, so surely not by 16, and the basic number is only divisible by 2 and not by further factors of 2.

Calculation modulo 16 is not difficult: every multiple of 400 is divisible by 16.

We see differences of 500 in table 5. A comparable kind of differences exists in the odd numbers: In table 3 indeed there are 8 squares ending on 121. The difference is in the thousands. The basic numbers 1, 7, 9 or 15 (mod 16) here give odd thousands, the numbers 3, 5, 11 or 13 (mod 16) squared result in even thousands.

This continues as far as one wants. Up to 10,000 there are 8 numbers which squared give four identical final digits. Try this out with e.g. 591, 1841, 3159, 4409, 5591, 6841, 8159 and 9409, all their squares end on 9281. Inevitably this symmetry is of paramount importance if we do integer roots.

Besides this symmetry we see more; in the even squares every number is 0 (mod 4) – zero modulo four, all the odd squares are 1 (mod 4), because 1^2 and 3^2 are both 1 (mod 4).

Besides not every number can be a square. Table 6 gives more insight (see next page).

So in fact there are 21 different possibilities in the squares up to 100 concerning the last two digits. Every number not mentioned in this table cannot be at the end of a square. Also there is the mention of the modulo 16 of the hundreds, which implies a restriction.

E.g. xx21396 cannot be a square, neither can 1825 et cetera et cetera.

Why is that? Divide 21,396 by 4 and you get 5,349, an odd hundred + 49 and therefore no square. In the case of hundred thousands, for each of them you can add 25,000 and the conclusion remains the same.

For 1825: according to the "fives trick" we have to multiply to consecutive numbers, 2 × 3, 47 × 48 or whatever and the result is a number ending on: 2, 6 or 0, so other possibilities drop out.

Table 5 shows that the squares are either 0 (mod 4) when the basic number is an even one. If the basic number is either 1 (mod 4) or 3 (mod 4), the squares of these numbers are without exception 1 (mod 4), according to table 4. So a sum of 2 different squares is either 0 (mod 4), 1 (mod 4) or 2 (mod 4) but never 3 (mod 4).

Table 5 shows us that all the squares of even numbers are 0 (mod 4) and the squares of the odd numbers are all 1 (mod 4). Regarding the sum of two squares this means that their sum can only be 0, 1 or 2 (mod 4), but never 3 (mod 4).

Table 6

Final digit	Tens	(H)undreds (E)ven / (O)dd
1	01	Always even, H all E (mod 16): 0, 2, 4, 6, 8, 10, 12, 14
	21	Always odd, H all O (mod 16): 1, 3, 5, 7, 9, 11, 13, 15
	41	Always even, H all E (mod 16): 0, 2, 4, 6, 8, 10, 12, 14
	61	Always odd, H all O (mod 16): 1, 3, 5, 7, 9, 11, 13, 15
	81	Always even, H all E (mod 16): 0, 2, 4, 6, 8, 10, 12, 14
4	04	E with H 0 (mod 4) and O with H 3 (mod 4)
	24	E with H 2 (mod 4) and O with H 3 (mod 4)
	44	E with H 6 and 14 (mod 16) and O with H 1 and 13 (mod 16)
	64	E with H 0 and 4 (mod 16) and O with H 1 and 9 (mod 16)
	84	E with H 4 and 12 (mod 16) and O with H 3 and 7 (mod 16)
5	25	Always even, H never 4 of 8, thousands 0 and 5
6	16	E with H 0 and 12 (mod 16), O with H 5 and 13 (mod 16)
	36	E with H 0 and 8 (mod 16) and O with H 3 and 15 (mod 16)
	56	E with H 2 and 6 (mod 16) and O with H 3 and 11 (mod 16)
	76	E with H 6 and 14 (mod 16) and O with H 5 and 9 (mod 16)
	96	E with H 8 and 12 (mod 16) and O with H 1 and 9 (mod 16)
9	09	Always even, H all E (mod 16): 0, 2, 4, 6, 8, 10, 12, 14
	29	Always odd, H all O (mod 16): 1, 3, 5, 7, 9, 11, 13, 15
	49	Always even, all E (mod 16): 0, 2, 4, 6, 8, 10, 12, 14
	69	Always odd, H all O (mod 16): 1, 3, 5, 7, 9, 11,13, 15
	89	Always even H, all E (mod 16): 0, 2, 4, 6, 8, 10, 12, 14

8.2 Creating squares

As in finding integer roots we can calculate the basic number, we can also create a square with the last digits we want to obtain using a variant of the well known $(a + b)^2 = a^2 + 2ab + b^2$. Let's say we want a square ending in 4969 with the basic number 13. $13^2 = 169$, so we want to add 4800. $2 \times 13 = 26$ and $48 \div 26 = 48$ or 98, we take 4813, and indeed $4813^2 = (2316) 4969$. We could also take as basic number $63^2 = 3969$. So now we have to add 1000. $2 \times 63 = (1)26$ and $1000 \div 26 = 35$, and indeed $3563^2 = (1269) 4969$. To complete the series: $5000 - 4813 = 187$, $5000 + 187 = 5187$, $10000 - 187 = 9813$, We get in order of magnitude: 187, 1437, 3563, 4813, 5187, 6437, 8563, 9813.

Also helpful is modulo 16 calculation. 969 ET (even thousand) is 9 (mod 16) so the

BN (basic number) will be 3, 5, 11 or 13 (mod 16). We can also find all the squares ending in 34969. We get in order of magnitude: 187, 6437, 43563, 49813, 50187, 56437, 93563, 99813.

Generally from the squares up to one thousand (and even higher) one can say: if you want to have a number of equal final figures in the squares, there are always as many possibilities as there are powers of 10.

If you want to have 8 squares ending on 687321, then start with $139^2 = 19321$. We are now missing 668000. We divide ?668 ÷ 139 and get 12, so the first square we look for is 12 ÷ 2 = 6, and $6139^2 = (37)\ 687321$. We now can find easily: 993861, 493861 and 506139. 687321 = 25 (mod 64), which means that the basic numbers are from: 5, 27, 37 or 59 mod 64. The ones we have are 6139 and 87611 = 59 mod 64 and their complements to 1,000,000, 912389 and 993861 are by consequence 5 mod 64. Now 500,000 minus 6139 = 493861, 37 mod 64. Then we get another 37 mod 64: 500,000 − 87611 = 412389. Now 1,000,000 − 412389 = 587611 and we have them all, in sequence: 6139, 87611, 412389. 493861, 506139, 587611, 912389 and finally 993861. The choice for mod 64 is because this is 2^6 as 1,000,000 is 10^6.

8.3 Squares, differences of

One thing is for sure: If a number is the difference between two different squares, then the number is not prime. After all, that's out of the question – here we find another well-known formula: $a^2 - b^2 = (a + b) \times (a - b)$.

The sport here is to find the factors, for which we have to find the squares involved.

Again we shall invoke the help of calculation modulo 8 and 9, because this spares us a lot of labour. The table will help us to sort out the numbers which we do not need.

The following abbreviations will be used:

- ET = even ten, such as 20, 40 et cetera
- OT = odd ten, such as 10, 30
- EH = even hundred, such as 00, 02 et cetera
- OH = odd hundred, such as 01, 03 et cetera
- DO = drops off

	Last two figures of given number	Smallest square	Biggest square
1	01, 21, 41, 61, 81	00 of 04, 24, 44, 64 of 84	01, 21, 41, 61, 81
2	03, 23, 43, 63, 83	Ending on 1	Ending on every 4, ET
3	05, 25, 45, 65, 85	Ending on 4	Ending on 9
4	07, 27, 47, 67, 87	Ending on 9	Ending on 6
5	09, 29, 49, 69, 89	Ending on 00 or 16, 36, 56, 76, 96	Ending on 00 or 25
6	11, 31, 51, 71, 91	Ending on 25 or 89, 69, 49, 29, 09	Ending on 16, 36, 56, 76, 96 of 00
7	13, 33, 53, 73, 93	Ending on 96, 76, 56, 36, 16	Ending on 09, 29, 49, 69, 89
8	15, 35, 55, 75, 95	Ending on 01, 21, 41, 61, 81	Ending on 16, 36, 56, 76, 96
9	17, 37, 57, 77, 97	Ending on 04, 24, 44, 64 84	Ending on 01, 21, 41, 61, 81
10	19, 39, 59, 79, 99	Ending on 01, 21, 41, 61, 81, 25	Ending on 04, 24, 44, 64, 84 of 00

Let's take as an example $817 = 7$ (mod 9), row 9 in the table. The smallest square to be added has to be 0 (mod 9) and end in a 4. After all, to obtain the sum of 7 (mod 9) the only possible combination is $0 + 7$ (mod 9). $17 + ?4$ has to result in a number ending in one and an even ten, other combinations are not possible. The square to be added has to be 0 (mod 9) and 0 (mod 4), so the basic number is a multiple of 6. We go to $12^2 = 144 = 0$ (mod 9) and indeed $817 + 144 = 961 = 31^2$.

So the answer is $817 = 31^2 - 12^2$.

3129: According to row 5 of the table. This number is 6 (mod 9) which means heaps of work! This is because it may be the difference between $1 - 4$ (mod 9), $4 - 7$ (mod 9) or $7 - 1$ (mod 9). If the smallest square ends in 96, the biggest one has to end in 25 and if the smallest one ends in 00 the biggest one ends in 29. So we have to complete 3129 with a square ending in 00 and which is 1, 4 or 7 (mod 9) with an even hundred or with a square ending in 96 with an even hundred and also 1, 4 or 7 (mod 9).

Hereunder follow the candidates and their reasons to drop off.

$14^2 = 196 = 7$ (mod 9) and OH and DO because of that: 3325 cannot be a square

$30^2 = 900 = $ OH and DO because of that: 4029 cannot be a square.

$36^2 = 1296 = $ EH, but 0 (mod 9). DO because a square cannot be 6 (mod 9), also 4425 cannot be a square, squares ending on 425 do not exist

$60^2 = 3600$, EH but 0 (mod 9) and squares 6 (mod 9) do not exist

$64^2 = 4096 = $ EH and 1 (mod 9) and $3129 + 4096 = 7225$, the square of 85.

So 3129 can be written as $85^2 - 64^2$ and we can factorise $85 + 64 = 149$ and $85 - 64 = 21$.

If we continue the factorisation we find $3129 = 3 \times 7 \times 149$. Another solution can be found on the base of 7×447 as follows:

$(447 \pm 7) \div 2 = 227$ and 220. We use this as follows:
$3129 + 220^2 = 3129 + 48400 = 51529 = 227^2$
Conclusion: $3129 = 227^2 - 220^2$
And $(1043 \pm 3) \div 2 = 523$ and 520.
Then we get $3129 + 520^2 = 523^2 = 3129 + 270400 = 273529$
In the sense of the question we write $3129 = 523^2 - 520^2$

14213: 2 (mod 9) row 7 of the table and 5 (mod 16). 14213 has an EH and the squares to be added on (49 and 89) are EH too. Or we have the squares ending in 16, 56 or 96 which have to be OH, as the squares ending on 09 always have EH and the squares ending on 29 and 69 always have OH. Moreover, the squares to be added have to be 7 (mod 9) to come to a square 0 (mod 9), And they have to be 4 (mod 16) to come to a square which is 9 (mod 16) as no square can be 5 (mod 16).

Hereunder follow the candidates and their reason to drop off (DO).

$04^2 = 16 = 7$ (mod 9) and DO because squares ending on 29 have OH.

$14^2 = 196$, 7 (mod 9), EH, DO because 14409 is not a square.

$76^2 = 5{,}776$, OH, 7 (mod 9), but 0 (mod 16) so DO.

$86^2 = 7{,}396$, OH and 4 (mod 16), let's take the risk: $14{,}213 + 7{,}396 = 21{,}609 = 147^2$. So $14{,}213 = 147^2 - 86^2$ and therefore we can see $14{,}213 = 61 \times 233$.

As the number 14213 only has two prime factors the above mentioned solution is the only one unless one uses $(14213 \pm 1) \div 2 = 7107^2 - 7106^2 = 14213$.

Because the squares are either 0 (mod 4) or 1 (mod 4), the difference between two different squares is either 0 (mod 4) or 3 (mod 4). There are no other possibilities. In the first mentioned case you can divide the number by four and then find the solution. And afterwards don't forget to multiply the solution found by two!

8.4   Squares, sums of 2

Questions of this nature are known by the theme of Pythagoras. Generally the sum of two squares is a "real square" in the sense that the sum of the two squares at the left side of the equals sign give a number which is exactly a square on the right side of the equals sign. We can even make questions of this nature ourselves. We chose the simplest form – the mathematic word is "primitive" – by taking an even and an odd number which have beside 1 no common divisor.

We take 3 and 2 which we call a and b. The formula for this activity is:

$(a^2 - b^2)^2 + (2ab)^2 = (a^2 + b^2)^2$

Worked out: $a^4 - 2a^2b^2 + b^4 + 4a^2b^2 = a^4 + 2a^2b^2 + b^4$

In numbers:

$(a^2 - b^2)^2 = (9 - 4)^2 = 5^2 = 25$
$(2ab)^2 = (2 \times 2 \times 3)^2 = 12^2 = 144$
$(a^2 + b^2)^2 = (9 + 4)^2 = 13^2 = 169$

And we see: $25 + 144 = 169$.

A practical adaption of the theme of Pythagoras is to be found in the construction business: if a right angle is desired, when making a wall, one takes a lath of 3 metres and one of four metres, joins the ends together with a lath of five metres and there is a right angle.

We could call a Pythagoras problem like this a "pure" Pythagoras and if we multiply each the elements of this by the same number there is a "derived" Pythagoras.

Questions: make "pure" Pythagoras with the numbers 4 and 7; 8 and 5; 13 and 6.

A little joke: Do you know the difference between a right angle and water? Solution by the answers.

After this we'll be engaged with the sum of two or more squares in general.

8.5 Sum of two squares

Talking about divisibility and factorisation we found that a number is divisible or can be factorised if it can be written as the sum of two squares in two different ways. The difference is that in the case of a division the divisor is known, in the case of factorisation we firstly have to find one factor before we can calculate the other one.

If a number can be written only once as the sum of two different squares, it is a prime number. More about this in the chapter "Factorisation".

We can spare ourselves a lot of work if we think about the possibilities before going to work. Are there numbers which cannot be the sum of two squares? The square of any even number is 0 (mod 4). The square of any odd number, 1 (mod 4) or 3 (mod 4), is 1 (mod 4). It is evident: if we add 0 + 1 (mod 4) the only result is 1 (mod 4). If we add 1 + 1 we get 2 (mod 4), so even numbers may be the sum of two squares. But a number 3 mod 4 can never be the sum of two (different) squares, as 1 + 1 does never exceed 2.

It is better is to invoke modulo 8 calculation to determine if a number is the sum of two squares, as we will have additions which exceed 100. An even number squared is either 0 (mod 8) or 4 (mod 8). Every odd number squared is 1 (mod 8). So the sum of two squares may be 0 (mod 8), 1 (mod 8), 4 (mod 8) or 5 (mod 8).

It is also useful to invoke modulo 9 calculation, as this will help us in economising the quantity of work we'll have to do. For the squares mod 9 there are the possibilities: 0, 1, 4 and 7 (mod 9). In sums of two squares there are the possibilities: 0, 1, 2, 4, 5, 7 and 8. 3 and 6 (mod 9) are impossible, as you can easily deduct.

We'll go and search.

Stimulated by my curiosity I made a table of two digit numbers which in theory can be the sum of two different squares. What we see is a total of 71 different numbers, the number of combinations as the sum of two different squares is 240.

It is evident that all the two digit numbers 3 (mod 4) dropped out. There are 71 numbers remaining from 00 up to 98.

Out of these 71 there are 11 numbers marked with a +, which means that these numbers are composed by the sum of two different squares and that this combination will always give an even number of hundreds. First example 18: squares ending on 09 have always an even hundred, and if you take 29, always odd hundred, you have to add 89, always even hundred, the result is 18 with an even hundred.

These events occur with numbers which, when divided by 2, result in a number 1 (mod 4). E.G: 06 divided by 2 gives 3, being 3 (mod 4), or 22 ÷ 2 = 11 = 3 (mod 4). All the other numbers can have either even or odd hundreds if they are the sum of two different squares.

The column E/O shows that the sum of the squares has an even hundred or an odd hundred. Nr Sq shows the number of sums of squares is given.

								Hundreds Even, (+) or Odd (−)				E/O	#Sq
00	=	00 + 00											1
00	=	04 + 96	24 + 76	44 + 56	64 + 36	84 + 16							5
01	=	00 + 01	25 + 76										2
05	=	01 + 04	21 + 84	41 + 64	61 + 44	81 + 24							5
05	=	16 + 89	36 + 69	56 + 49	76 + 29	96 + 09							5
06	=	25 + 81										−	1
08	=	04 + 04	24 + 84	44 + 64									3
09	=	00 + 09	25 + 84										2
10	=	01 + 09	21 + 89	41 + 69	61 + 49	81 + 29							5
12	=	16 + 96	36 + 76	56 + 56	+	+							3
13	=	04 + 09	24 + 89	44 + 69	64 + 49	84 + 29							5
14	=	25 + 89										−	1
16	=	00 + 16											1
17	=	01 + 16	21 + 96	41 + 76	61 + 56	81 + 36							5
18	=	09 + 09	29 + 89	49 + 69								+	3
20	=	04 + 16	24 + 96	44 + 76	64 + 56	84 + 36							5
21	=	00 + 21	25 + 96									−	2
22	=	01 + 21	41 + 81	61 + 61									2
24	=	00 + 24											3
25	=	00 + 25	04 + 21	41 + 84	61 + 64	81 + 44						+	5
26	=	01 + 25										+	1
28	=	04 + 24	44 + 84	64 + 64									3
29	=	00 + 29	04 + 25										2
30	=	01 + 29	21 + 09	41 + 89	61 + 69	81 + 49						−	5
32	=	16 + 16	36 + 96	56 + 76									3
33	=	04 + 29	24 + 09	44 + 89	64 + 69	84 + 49							5
34	=	09 + 25										+	1
36	=	00 + 36											1
37	=	01 + 36	21 + 16	41 + 96	61 + 76	41 + 96							5
38	=	09 + 29	49 + 89	69 + 69	+	+						−	3
40	=	04 + 36	24 + 16	44 + 96	64 + 76	84 + 56							5
41	=	00 + 41	25 + 16										2
42	=	01 + 41	21 + 21	61 + 81								+	3
44	=	00 + 44											1
45	=	01 + 44	21 + 24	41 + 04	61 + 84	81 + 64							5
45	=	09 + 36	29 + 16	49 + 96	69 + 76	89 + 56							1
46	=	25 + 21										−	1
48	=	04 + 44	24 + 24	64 + 84									3
36											Total		115

8.5 Sum of two squares

											Hundreds Even, (+) or Odd (-)					E/O	# Sq
49	=	00 + 49	25 + 24														2
50	=	01 + 49	21 + 29	41 + 09	61 + 89	81 + 69										+	5
50	=	25 + 25															1
52	=	16 + 36	56 + 96	76 + 76													3
53	=	04 + 49	24 + 29	44 + 09	64 + 89	84 + 69											5
56	=	00 + 56															1
57	=	01 + 56	21 + 36	41 + 16	61 + 96	81 + 76											5
58	=	09 + 49	29 + 29	69 + 89												+	3
60	=	04 + 56	24 + 36	44 + 16	64 + 96	84 + 76											5
61	=	00 + 61	25 + 36														2
62	=	01 + 61	21 + 41	81 + 81												-	3
64	=	00 + 64															1
65	=	01 + 64	21 + 44	41 + 24	61 + 04	84 + 81											5
65	=	16 + 49	36 + 29	56 + 09	76 + 89	96 + 69											5
66	=	41 + 25															+	1
68	=	04 + 64	24 + 44	84 + 84													3
69	=	00 + 69	25 + 44														2
70	=	01 + 69	21 + 49	41 + 29	61 + 09	81 + 89										-	5
72	=	16 + 56	36 + 36	76 + 96	+	+											3
73	=	04 + 69	24 + 49	44 + 29	64 + 09	64 + 89											5
74	=	25 + 49															+	1
76	=	00 + 76															1
77	=	01 + 76	21 + 56	41 + 36	61 + 16	81 + 96											5
78	=	09 + 69	29 + 49	89 + 89	+	+											3
80	=	04 + 76	24 + 56	44 + 36	64 + 16	84 + 96											5
81	=	00 + 81	25 + 56														2
82	=	01 + 81	21 + 61	41 + 41													3
84	=	00 + 84															1
85	=	01 + 84	21 + 64	41 + 44	61 + 24	81 + 04											5
85	=	16 + 69	36 + 49	56 + 29	76 + 09	96 + 89											5
86	=	25 + 61															-	1
88	=	04 + 84	24 + 64	44 + 44													3
89	=	00 + 89	25 + 64														2
90	=	01 + 89	21 + 69	41 + 49	61 + 29	81 + 09										+	5
92	=	16 + 76	36 + 56	96 + 96	+	+											3
93	=	04 + 89	24 + 69	44 + 49	64 + 29	84 + 09											5
94	=	25 + 69															-	1
96	=	00 + 96															1
97	=	01 + 96	21 + 76	41 + 56	61 + 36	81 + 16											5
98	=	09 + 89	29 + 69	49 + 49												+	3

37   Total   125

8.6 Congrua

In the book "Prime numbers", written by David Wells (and edited in 2005 by John Wiley and sons, Hoboken, New Jersey) we find on page 101 a question which was presented to Fibonacci as follows: find a square which remains a square if 5 is added or subtracted. The answer, in fractions, is:

$$\left(\frac{41}{12}\right)^2 + 5 = \left(\frac{49}{12}\right)^2 \text{ and } \left(\frac{41}{12}\right)^2 - 5 = \left(\frac{31}{12}\right)^2$$

This was rather artificial in my feeling and I multiplied it all by 144 and came up with:

- $41^2 - 5 \times 12^2 = 31^2$
- $41^2 + 5 \times 12^2 = 49^2$

After that my question was this: Is there a square which remains a square if one adds a number and remains a square if this number is subtracted? And: Can a formula be found for this?

After some seeking I found:

- $5^2 - 24 = 1^2$ and $5^2 = 25 + 24 = 7^2 = 49$
- $13^2 - 120 = 7^2 = 49$ and $13^2 = 169 + 120 = 17^2 = 289$
- $101^2 - 3960 = 79^2 = 6241$ and $101^2 = 10201 + 3960 = 119^2 = 14161$

What is happening here? The difference between 5^2 and 1^2 can be explained by $(a + b) \times (a - b)$: $(5 + 1) \times (5 - 1) = 6 \times 4 = 24$. And $7^2 - 5^2 = (7 + 5) \times (7 - 5) = 2 \times 12 =$ also 24.

One can do the same with $13^2 - 7^2 = (13 + 7) \times (13 - 7) = 20 \times 6 = 120$. And $17^2 - 13^2 = (17 + 13) \times (17 - 13) = 30 \times 4 =$ also 120.

What I found was that all the numbers are multiples of 24 with, for example, 18 or 36 where the combinations cannot be made. There has to be a formula for that, but which one?

Dr. Benne de Weger, a mathematician at the university in Eindhoven, gave the solution. You'll find his name more often in this book. This way of working with squares is known as congruüm – plural congrua – the title of this chapter, and here follows the explanation. To construct such a congruüm in its most simple form – in mathematics they say "primitive" – we take two numbers, m and n, one even and one odd, which beside 1 do not have a common divisor.

This has the following consequences:

- If n is even and m is odd, and we square $(m + n)^2$ we get a number which is 1 (mod 4).
- If neither m nor n are divisible by three, this implies that the differences

between the squares are divisible by three, so the difference is 0 (mod 3).
- The difference between two odd squares is invariably 0 (mod 8).

Because of these two facts it can be concluded that a congruüm is always divisible by 24.

We now start with m = 3 and n = 2 and look for the numbers a, b and c, which being squared, form a congruüm.

A congruüm can be calculated using the formula $4 \times m \times n \times (m + n) \times (m - n)$.

Now we see:
- $a = (m + n)^2 - 2m^2 = (3 + 2)^2 = 5^2 = 25 - (2m^2) = 25 - 18 = 7$
- $b = (m + n)^2 - 2n^2 = (3 + 2)^2 = 5^2 - (2n^2) = 25 - 8 = 17$
- $c = m^2 + n^2 = 3^2 + 2^2 = 13$

The congruüm is $4 \times 3 \times 2 \times (3 + 2) \times (3 - 2) = 120$

The proof: $13^2 = 169 - 120 = 49 = 7^2$ and $13^2 = 169 + 120 = 289 = 17^2$

The second work out is with m = 10 and n = 7 which gives:
- $a = (m + n)^2 - 2m^2 = (10 + 7)^2 - 2 \times 10^2 = 17^2 = 289 - 2 \times 100 = 289 - 200 = 89$
- $b = (m + n)^2 - 2n^2 = (10 + 7)^2 = 17^2 = 289 - 2 \times 49 = 289 - 98 = 191$
- $c = m^2 + n^2 = 10^2 + 7^2 = 149$.

The congruüm is $4 \times 10 \times 7 \times (10 + 7) \times (10 - 7) = 14280$

The proof: $149^2 = 22201 - 14280 = 7921 = 89^2$ and $149^2 = 22201 + 14280 = 36481 = 191^2$.

So the basic numbers are a = 89, b = 191 and c = 149.

8.6.1 Which squares?

On the internet circulates a list with numbers which are described as congrua. Now it is interesting to find out which squares are hidden behind these numbers.

These are some of them: 120, 216, 240, 336, 384, 480, 600, 720, 840, 864, 960, 1080, 1320, 1344, 1536, 1920, 1944, 2016, 2160, 2184, 2400, 2520, 2880, 3000, 3024, 3360, 3456, 3696, 3840, 3960, 4056, 4320, 4704, 5376, 5400, 5544.

We will not discuss all these numbers, but we will trace if with the aid of the formula the squares and next the basic numbers can be retrieved, if the numbers are primitive or composed. Our first look is at the numbers 120, 480 and 1920; and 216, 864 and 3456; 336 and 1344 and 5376 which differ by a factor of 4. There are in the list also numbers which differ by a factor of 9 such as: 216 and 1944; 480 and 4320; and the number 4056 is 24 times 169.

Apart from that, nothing forbidden has taken place; nowhere is there prescribed that a congruüm should have a primitive form.

The first number which awakened my suspicion was 600. Based on the formula it is divided by 4, with the result 150. We write down all the divisors of 150: 1, 2, 3, 5, 6, 10, 15, 25, 30, 50, 75, 150.

And we factorise 150: $2 \times 3 \times 5 \times 5$.

Now the question is: Can we split up 150 in such a way that the result fits in the formula: $m \times n \times (m + n) \times (m - n)$? We can restrict ourselves to the lower numbers; after all there is a sum in it of two numbers and the product has to be 150.

You may try as much as you want, but there is no solution. How did the number 600 appear in the list? Well, divide the number by 25 and you get 24 and this fits perfectly in the squares of 1, 5 and 7: 1, 25 and 49. 1, 5 and 7 multiplied by 5 gives 5, 25 and 35 and their respective squares are 25, 625 and 1225, and look, there is our 600. And if we divide 864 by 36 we get again 24 with exactly the same basic numbers. In both cases there are derived congrua.

The same is valid for 4704: if you divide that by 196 you get 24.

We now take $3696 \div 4 = 924$. The factorisation is $2 \times 2 \times 3 \times 7 \times 11$. The divisors are: 1, 2, 3, 4, 6, 7, 11, 12, 14, 21, 28, 33, 44, 66, 77, 132, 154, 231, 308, 462, 924. This is for the completeness; we can restrict ourselves to 14 as the biggest divisor.

The formula: We have already divided 3696 by four: $m \times n \times (m + n) \times (m - n)$. The question: can we factorise 924 in such a way that it fits in the formula? One of the factors has to be a 7 and there also has to be an 11. It lies at hand that we will work with 7 and 4 as $7 + 4 = 11$ and $7 - 4 = 3$. So $m = 7$ and $n = 4$.

The work out: $7 \times 4 \times (7 + 4) \times (7 - 4) = 7 \times 4 \times 11 \times 3 = 924$. The basic numbers for the squares will be then: $(m + n)^2 - 2 m^2 = 11^2 - 2 \times 7^2 = 121 - 98 = 23$ as a and $(m + n)^2 - 2n^2 = 11^2 - 2 \times 4^2 = 121 - 32 = 89$ as b. For $c = 7^2 + 4^2 = 65$. The squares are respectively 529, 4225 and 7921 and the congruüm is 3696.

The basic numbers are $a = 23$, $b = 89$ en $c = 65$.

Because it is such a nice sport we do another one: $2184 \div 4 = 546$. The question: can we factorise 546 in such a way that it fits in our formula $m \times n \times (m + n) \times (m - n)$? Factorising gives this: $2 \times 3 \times 7 \times 13$. Some divisors of 546 are 1, 2, 3, 6, 7, 13, 14. If we have a look at this the answer is presented to us: $6 \times 7 \times 13$ is 546 and so we get $m = 7$ and $n = 6$, and by consequence $(m + n) = 13$ and $(m - n) = 1$.

Then the basic numbers are: 71, 85 and 97.

$a = (m + n)^2 - 2m^2 = 13^2 - 2 \times 7^2 = 169 - 98 = 71; 71^2 = 5041$

$b = (m + n)^2 - 2n^2 = 13^2 - 2 \times 6^2 = 169 - 72 = 97, 97^2 = 9409$

$c = m^2 + n^2 = 49 + 36 = 85; 85^2 = 7225$

Then $7225 - 5041 = 2184$ and $9409 - 7225 = $, and you guess already also 2184, our congruüm.

8.7 Answers to the Pythagorases

If $a = 7$ and $b = 4$ then we get $a^2 - b^2 = 49 - 16 = 33$, $2ab = 2 \times 4 \times 7 = 56$ and $a^2 + b^2 = 49 + 16 = 65$. The answer is $33^2 + 56^2 = 65^2$ and that is – of course – correct as $1089 + 3136 = 4225$.

If $a = 8$ and $b = 5$ then we have $a^2 - b^2 = 64 - 25 = 39$; $2ab = 2 \times 8 \times 5 = 80$ and $a^2 + b^2 = 89$. The answer is $39^2 + 80^2 = 89^2$ and we see this: $1521 + 6400 = 7921$.

If $a = 13$ and $b = 6$ then is $a^2 - b^2 = 169 - 36 = 133$; $2ab = 2 \times 13 \times 6 = 156$ and $a^2 + b^2 = 169 + 36 = 205$. The answer is $133^2 + 156^2 = 205^2$, and also this is correct as $17689 + 24336 = 42025$.

The little joke: water cooks at 100° and a right angle cooks at 90°.

8.8 More about sums of squares

8.8.1 Sums of squares in relation to divisibility

Whoever is confident with the sums of squares can find herein a very useful expedient for division or factorisation. The difference is that in the case of a division we know at least one divisor, but in the case of factorisation we do not know anything. The given number has to be divided, but by which number?

Examples

$$65$$

This is the smallest possible one. This number can be written as $8^2 + 1^2$ or $7^2 + 4^2$. We take the even numbers and the odd numbers apart and calculate $(8 \pm 4) \div 2$ and get respectively 6 and 2. Then $(7 \pm 1) \div 2$ will give us 4 and 3. Next we take 6 and 3 and calculate their common factor, 3 and the number of times, here 2× and 1×. Then we take 4 and 2 and calculate their common factor, 2, respectively 2× and 1×. Next we take the two common factors, 3 and 2 and square them and add the squares. Then we have $3^2 + 2^2 = 13$, which is one of the factors of 65. The final operation is that we take the number of times, in this case respectively 2 and 1, then square these numbers and add the squares. We get $2^2 + 1^2 = 5$ and now we have found the second factor of 65, which is 5. So the factorisation of $65 = 5 \times 13$.

Also, we take the even basic numbers of the squares, 8 and 4 and do $(8 \pm 4) \div 2$ and we get 6 and 2. The odd basic numbers $(7 \pm 1) \div 2$ give us 4 and 3. Now multiply both answers and look: $6 \times 2 = 4 \times 3$.

377

To find the composing squares means a lot of work. 77 can be composed by a lot of numbers: 01 + 76; 21 + 56; 41+ 36; 61 + 16; 81 + 96, no fewer than five different possibilities. We do a smart thing: we double the number and get 54 and now we have our advantage, as this number can only be composed by 29 + 25.
So we continue with $2 \times 377 = 754$. And unexpectedly we have a second advantage: 754 is 7 mod 9 which means that one of the composing squares will be 0 mod 9, so one basic number is divisible by 3 and the other one is 4 or 5 mod 9. Also one of the basic numbers is a multiple of 5.
As 754 is destining we may assume that the 5 numbers candidate is either 5, as its square is 7 mod 9 or 15, as its square is 0 mod 9. 25 drops off as its square 625 which is 4 mod 9.

Concerning the 29 there are two candidates: 23 as its square 529 is 7 mod 9, and 27 as its square 27 is 0 mod 9.

By coincidence the numbers mentioned are also the numbers we look for: 754 is the sum of both $23^2 + 15^2$ and $27^2 + 5^2$. Next step: $(27 \pm 23) \div 2 = 25$ and 2. Then $(15 \pm 5) \div 2 = 10$ and 5. Next 25 and 10 have the common factor 5, respectively 5× and 2×. And 5 and 2 have common factor 1, respectively 5× and 2×. One of the factors is $5^2 + 2^2 = 29$. The other one is $5^2 + 1^2$, and here we make a correction. Remember, we doubled the number 377 but we factorised 754, and now we correct this by attending to both odd factors 5 and 1 via $(5 \pm 1) \div 2 = 3$ and 2. By squaring and adding $3^2 + 2^2 = 13$ we have found the second factor, so $377 = 13 \times 29$.

33973

Again we have a number with multiple combinations: 09 + 64; 29 + 44; 49 + 24; 69 + 04; 89 + 84. Again we double and get 67946. This number is 5 mod 9 which means that the mod 9 combinations can be only 1 + 4 or 7 + 7. Neither basic number is divisible by 3 as then the square is 0 mod 9 which we cannot combine with another square to a sum which is 5 mod 9. The numbers we look for have to end in 21 and 25, there are no other possibilities.

We have here a table of numbers which square ends in 21 and give the reason why it is either a candidate or drops out.

B.N. means basic number.

Look how interesting this is! By means of elimination we can immediately skip 7 out of 10 numbers with squares which end in 21.

B.N.	Square	Drops out	B.N.	Square	Drops out	B.N.	Square	Drops out
11	121	No sq. 825	111	12321	0 mod 9	211	44521	No sq. 425
39	1521	No sq .425	139	19321	No sq. 8625	239	57121	No sq. 825
61	3721	Cand.	161	25921	Cand.			
89	7921	Cand.	189	35721	0 mod 9			

As no square exists ending in 425 or 825, four squares drop out. And 625 only can be a square if the thousand is 0 or 5.

61^2 does not fit, because 64225 is no square. 89^2 fits, as $69746 - 7921 = 60025$, the square of 245. Our first combination is $245^2 + 89^2$. $67946 - 25921 = 42025$, the square of 205. Our second combination is $67946 = 205^2 + 161^2$.

To find our factors/divisors we do this: $(245 \pm 205) \div 2 = 225$ and 20. And $(161 \pm 89) \div 2 = 125$ and 36. The common factor of 225 and 125 is 25, respectively 9× and 5×. The common factor of 36 and 20 is 4, respectively 9× and 5×. Next we take the both common factors, and as one of them is even and one of them is odd, we square them and add the squares, and the result is an odd number which is just what we want. Then we get $25^2 + 4^2 = 641$. The number of times (above) is 9 and 5, which we need to correct as we doubled the question number. Now we take $(9 \pm 5) \div 2$ and get 7 and 2. We square these and add the squares, the result being 53. We have now completed our division/ factorisation, the result is 641×53.

38641

If the last two figures of a number can be the last two figures of a square, then there are always more possibilities for the sums of squares. In this case for a number ending in 41 this means that the sum of the composing squares can be either 00 + 41 or 25 + 16.

The number is 4 mod 9, which means that one of the squares will be 0 mod 9, so the basic number is a multiple of 3 and the other one is 4 mod 9, so the basic number is 2 or 7 mod 9.

As the squares ending on 41 always have an even hundred, the squares we look for must also have an even hundred. Following we have two tables, one for the even hundreds and one for the 41s.

B.N.	Square	Drops out	B.N.	Square	Drops out
20	400	Cand.	120	14400	Cand.
40	1600	7 mod 9	140	19600	7 mod 9
60	3600	Cand.	160	25600	Cand.
80	6400	1 mod 9	180	32400	Cand.
100	10000	1 mod 9	200	40000	Exceeds

Here we see that in our first selection four numbers drop out.

B.N.	Square	Drops out	B.N.	Square	Drops out
21	441	38200 no square	121	14641	7 mod 9
29	841	37800 no square	129	16641	22000 no square
71	5041	1 mod 9	171	29241	9400 no square
79	6241	Cand.	179	32041	1 mod 9

About the hundreds: 20^2 drops out, as 38241 is no square, 120^2 drops out because 24241 is no square, 160^2 drops out as 13041 is no square, 180 is a hit, as $38641 - 32400 = 6241$, which is the square of 79. So now we have $38641 = 180^2 + 79^2$.

About the combinations 25 + 16. Here we have the same choice: the combinations of 25 and 16 and the 0 mod 9 + 4 mod 9. As our question number has an even hundred, the 16 candidates should each also have an even hundred, as all the squares ending in 25 have always and automatically an even hundred.

B.N.	Square	Drops out	B.N.	Square	Drops out	B.N.	Square	Drops out
5	25	7 mod 9	105	11025	27616 no sq.	4	16	7 mod 9
15	225	Cand.	115	13225	25416 no sq.	96	9216	29425 no sq.
25	625	38016 no sq.	125	15625	1 mod 9	104	10816	7 mod 9
35	1225	1 mod 9	135	18225	20416 no sq.	196	38416	Cand.
45	2025	36616 no sq.	145	21025	1 mod 9			
55	3025	1 mod 9	155	24025	14616 no sq.			
65	4225	34416 no sq.	165	27225	11416 no sq.			
75	5625	33016 no sq.	175	30625	7 mod 9			
85	7225	7 mod 9	185	34225	7 mod 9			
95	9025	7 mod 9	195	38025	616 no sq.			

Looking over this table we see that in fact there is, both in the twenty-fives and the sixteens, only one combination; $15^2 + 196^2$. We had already seen $180^2 + 79^2$ and now we look at $(79 \pm 15) \div 2 = 47$ and 32. Next comes $(196 \pm 180) \div 2 = 188$ and 8. The common factor of 188 and 47 is 47, respectively 4× and 1×. The common factor of 32 and 8 is 8, respectively 4× and 1×. So the factors for 38641 are $47^2 + 8^2 = 2273$ and $4^2 + 1^2 = 17$.

8.8.2 Sums of squares and factoring

There is (much) more to be said about the sums of squares in combination with factoring.

What happens if one of the factors itself is a square or if one of the factors is a third or even fourth power?

841

This number is 29^2 which we can only compose by adding $20^2 + 21^2$. As zeros do not make sense compositions as $29^2 + 0^2$ will be ignored.

125

This number is 5^3 which we can compose by adding $11^2 + 2^2$ and $10^2 + 5^2$. In the case of a number which is cubed, we see that the basic number appears one time in the addition of the squares.

2197

This number is 13^3 can be composed by adding $39^2 + 26^2$ (both multiples of 13), and

$46^2 + 9^2$. There are no more possibilities.

325

Here we take the square of five in combination with 13. The sums of squares are: $10^2 + 15^2$; $17^2 + 6^2$; $18^2 + 1$. Now we have 3 pairs of sums of squares. Aha, one of the factors is squared which indicates that the concerning factor will appear in the row of sums of squares. For finding the factors we take $15^2 + 10^2$ and $18^2 + 1^2$. Next $(18 \pm 10) \div 2 = 14$ and 4, then $(15 \pm 1) \div 2 = 8$ and 7. The common factor of 14 and 7 is 7, respectively 2× and 1×. The common factor of 8 and 4 is 4, respectively 2× and 1×. The factors of 325 then are $7^2 + 4^2 = 65$ and $2^2 + 1^2 = 5$.

What will happen if one of the factors has a third power in it? We now take 5^3. And see, we have:

1625

These are the sums of squares: $28^2 + 29^2$; $35^2 + 20^2$; $37^2 + 16^2$; $40^2 + 5^2$. Here we have four combinations in the sums of squares. We see immediately that now the factor which there is a third power appears twice in the sum of squares: $35^2 + 20^2$ and $40^2 + 5^2$. The method for finding the factors remains always the same; we now do $28^2 + 29^2$ and $40^2 + 5^2$ and get $(29 \pm 5) \div 2 = 17$ and 12. Then we have $(40 \pm 28) \div 2 = 34$ and 6. The common factor of 34 and 17 is 17, respectively 2× and 1×. The common factor of 12 and 6 = 6, respectively 2× and 1×. The factors of 1625 then are $17^2 + 6^2 = 325$ and $2^2 + 1^2 = 5$. We can also take the combinations $35^2 + 20^2$ and $37^2 + 16^2$ and then get $(37 \pm 35) \div 2$, which results in 36 and 1. Next we have $(20 \pm 16) \div 2$ gives 18 and 2. 36 and 18 have common factor 18, respectively 2× and 1×. 2 and 1 have common factor 1, respectively 2× and 1×. So the factors of 1625 are now $18^2 + 1^2 = 325$ and $2^2 + 1^2 = 5$.

We take a fifth power and get

40625

As sums of squares, $200^2 + 25^2$; $199^2 + 32^2$; $196^2 + 47^2$; $185^2 + 80^2$; $175^2 + 100^2$; $145^2 + 140^2$. Now we have 6 combinations of which $199^2 + 32^2$ and $196^2 + 47^2$ have the simplest form and the other combinations are all divisible by 5.

203125

You will recognise this as $5^6 \times 13$, and this sixth power gives us seven different combinations. They are $450^2 + 25^2$; $439^2 + 102^2$; $430^2 + 135^2$; $425^2 + 150^2$; $375^2 + 250^2$; $366^2 + 263^2$; $345^2 + 290^2$. As you see, only $439^2 + 102^2$ and $366^2 + 263^2$ are simple ones, the other combinations are derivatives, and are divisible by 5.

1015625

This is $5^7 \times 13$ and results in the following eight combinations: $1000 + 125^2$; $995^2 + 160^2$; $980^2 + 235^2$; $925^2 + 400^2$; $892^2 + 469^2$; $875^2 + 500^2$; $776^2 + 643^2$ and $725^2 + 700^2$. Here also we see only two simple combinations and all the other ones are derived.

Finally the product of two fourth powers; for the results here I invoked the help of my computer.

17850625

which you get if you calculate 65^4. Here we have 12 different combinations, of which only 2 are simple ones. In the case of composed combinations the common factor is given between brackets.

The combinations are $4199^2 + 468^2$ (common factor 13); $4185^2 + 580^2$ (common factor 5); $4180^2 + 615^2$ (common factor 5); $4095^2 + 1040^2$ (common factor 13); $4056^2 + 1183^2$ (common factor 13); $3900^2 + 1625^2$ (common factor 65); $3713^2 + 2016^2$ (no common factor); $3696^2 + 2047^2$ (no common factor); $3640^2 + 2145^2$ (common factor 65); $3289^2 + 2652^2$ (common factor 13); $3000^2 + 2975^2$ (common factor 25).

Finally: all these interesting things can exclusively be done with prime numbers 1 mod 4, so the sum of two different squares, one even, one odd and no common divisor.

Prime numbers 3 mod 4 do not offer these possibilities.

8.9   Sums of three squares

Of course sums of three squares offer a wider range of possibilities, however not everything is possible. Let's remember from the structure in the squares: without any exception there are only two possibilities: a square is either 0 (mod 4) or 1 (mod 4). And we concluded: a number 3 (mod 4) can never be the sum of two different squares.

As an introduction to the theme:

632623, can this number be the sum of three different squares?
By coincidence I saw this number on a van, as the phone number of a company.

8.9.1 Arithmetical solution

We can obtain 23 as end-digits by:

1. Adding the squares which end in 01, 41 and 81. They all have an even hundred and by consequence the sum of them to get 23 as final digits will have an odd hundred, so this is impossible.
2. Adding the squares ending in 21 and adding two times 01 gives 23. But with an odd hundred as the 21s have and two times an 01 with always an even hundred, this results always in 23 with an odd hundred, so this is also impossible.
3. Adding 2 squares ending in 61, each with always an odd hundred, gives 22 with an odd hundred and further on adding 01 with an even hundred results in 23 with an odd hundred. Again, impossible.
4. In combination with 3 squares ending on 09, 49 and 89, all always even hundreds, or ending on 29 and 69, all always odd hundreds, 77 can be composed, but 23 cannot.
5. Two combinations 24 give 48, but 75 is never a square.
6. For getting 23 another 50 is needed which is never a square.
7. Conclusion: we cannot obtain the sum we want.

8.9.2 Algebraic solution

Now we use modulo calculation. As every 1,000 is divisible by 8, we only need to consider 623, which is 7 (mod 8). For all the squares up to 8 there are these possibilities: they can be 0, 1 or 4 (mod 8). In case of an addition of three squares we can obtain the following mod 8 remainders: 0, 1, 2, 3, 5, and 6 mod 8, but never 7 mod 8.

Conclusion therefore: 632623 cannot be the sum of three squares.

8.9.3 The numbers

For studying the sum of three different squares we take the same numbers as in the chapter "sum of four different squares".

For the good order: if you see three possibilities mentioned, certainly you can find more. It was not my intention to list them all.

For finding eventual difficulties I took randomly the number 15120 and lower for looking after the sum of three different squares. My aim is to indicate the problems you'll meet, I do not go down to 7560, the half of 15120. In the range of numbers you'll see that most of the problems will be mentioned.

15120: $120^2 + 24^2 + 12^2$, $121^2 + 479$, 7 (mod 8), so impossible to be the sum of two squares; $119^2 + 959$ is impossible as the sum of two squares because $959 = 7$ (mod 8), $118^2 + 1196 \div 4 = 299 = 3$ (mod 4) so impossible as the sum of two squares, $116^2 + 1664 = 8^2 + 40^2$, a good solution. $114^2 + 2124 = 6^2 \times 59 = 3$ (mod 4) and thus not the sum of two squares. Also $2124 \div 4 = 531 = 3$ (mod 4) so not the sum of two squares. $112^2 + 2576$, again not the sum of two squares. $2576 \div 16 = 161$ which is not the sum of two squares. $110^2 + 3020$, not the sum of two squares, as $3020 \div 4 = 755$, which is 3 (mod 4). 108, 106, 104 and 102 are also impossible, and remaining is: $32^2 + 64^2 + 100^2$.

15119: This number is 7 (mod 8) so that no further attempts need to be made.

15118: $122^2 + 234 = 3^2 + 15^2$; $121^2 + 477 = 6^2 + 21^2$, $120^2 + 718 \div 2 = 359$, 3 (mod 4), not the sum of two squares, $119^2 + 957 \div 3 = 319$, 3 (mod 4), not the sum of two squares, $15118 \div 2 = 7559$, 3 (mod 4), not the sum of two squares, $117^2 + 1429 = 23^2 + 30^2$; $114^2 + 2122 = 41^2 + 21^2$; $113^2 + 45^2 + 18^2$. As 15118 is 7 (mod 9) a solution should have two squares 0 (mod 9) and one 7 (mod 9).

15117: $122^2 + 233 = 8^2 + 13^2$; $121^2 + 476$, $\div 4 = 119$, 3 (mod 4), not the sum of two squares; $120^2 + 717 \div 3 = 239$, 3 (mod 4), not the sum of two squares, $119^2 + 956 \div 4 = 239$, 3 (mod 4), not the sum of two squares, $118^2 + 1193 = 13^2 + 32^2$, a lot more impossibilities, then $106^2 + 3881 = 20^2 + 59^2$. As $15117 = 6$ (mod 9) no squares 0 (mod 3) are possible. Another good solution is $101^2 + 4916 = 4^2 + 70^2$.

15116: $122^2 + 232$, $6^2 + 14^2$; $121^2 + 475 = 3$ (mod 4), not the sum of two squares, $120^2 + 716 \div 4 = 179 = 3$ (mod 4), not the sum of two squares; only even squares have a chance, as the odd ones give 3 (mod 4), not the sum of two squares. $118^2 + 1192 = 6^2 + 34^2$, $116^2 + 1660 \div 4 = 415$, 3 (mod 4), not the sum of two squares; $114^2 + 2120 = 2^2 + 46^2$, $26^2 + 38^2$, $112^2 + 2572 \div 4 = 643$, 3 (mod 4), not the sum of two squares, $110^2 + 3016 = 10^2 + 54^2$ and $30^2 + 46^2$

15115: $122^2 + 231 = 3$ (mod 4), not the sum of two squares, $121^2 + (474 \div 6) = 79$, 3 (mod 4), not the sum of two squares, $120^2 + 715 = 3$ (mod 4), not the sum of two squares; $119^2 + 954 = 15^2 + 27^2$, as $15115 = 4$ (mod 9), only subtract an odd square 0 (mod 9): $99^2 + 5314 (= 33^3 + 65^2)$. And $93^2 + 6466 (= 79^2 + 15^2)$ and $75^2 + 29^2$.

15114: $122^2 + 230 \div 2 = 115$, 3 (mod 4), not the sum of two squares, $121^2 + 473$, not the sum of two squares, $120^2 + 714$, not the sum of two squares, $119^2 + 953 = 13^2 + 28^2$, $118^2 + 1190 \div 2 = 595$, 3 (mod 4), not the sum of two squares. A solution can only be found by subtracting an odd square $\neq 0$ (mod 9). Example: $15114 - 91^2 = 6833 = 68^2 + 47^2$.

15113: $122^2 + 229 = 2^2 + 15^2$, $121^2 + 472 \div 8 = 59$, 3 (mod 4), not the sum of two squares, $120^2 + 713$, not the sum of two squares, $119^2 + 952 \div 8 = 119$, 3 (mod 4), not the sum of two squares, $118^2 + 1189 = 10^2 + 33^2$, $17^2 + 30^2$, $117^2 + 1424 = 20^2 + 32^2$, $116^2 + 1657 = 19^2 + 36^2$, $115^2 + 1888 \div 32 = 59$, 3 (mod 4), not the sum of two squares, $114^2 + 2117 = 1^2 + 46^2$, $31^2 + 34^2$.

15112: $122^2 + 228$ = impossible; $121^2 + 471 = 3$ (mod 4) and so not the sum of two squares, $120^2 + 712 = 6^2 + 26^2$, $119^2 + 951 = 3$ (mod 4), not the sum of two squares, $118^2 + 1188$, impossible, $117^2 + 1423 = 3$ (mod 4), not the sum of two squares, $116^2 + 1656 \div 8 = 207$, 3 (mod 4), not the sum of two squares, $115^2 + 1887$, 3 (mod 4), not the sum of two squares, $114^2 + 2116 = 46^2$, not the sum of two squares, this number is 1 (mod 9), so subtracting an even square could help: $9604 + 5508 = 18^2 + 72^2$, $6084 + 9028 = 48^2 + 82^2$ (or $62^2 + 72^2$).

15111: whatever we do, there is no solution for this number as it is 7 (mod 8).

15110: $122^2 + 226 = 1^2 + 15^2$, $121^2 + 469 = 7 \times 67$, 3 (mod 4), both not the sum of two squares, $120^2 + 710 \div 2 = 355$, 3 (mod 4), not the sum of two squares, $119^2 + 949 = 18^2 + 25^2$, $7^2 + 30^2$, $118^2 + 1186 = 15^2 + 31^2$, $117^2 + 1421 = 14^2 + 35^2$, $116^2 + 1654 \div 2 = 827$, 3 (mod 4), not the sum of two squares, $115^2 + 1885 = 6^2 + 43^2$, $11^2 + 42^2$, $27^2 + 34^2$.

15109: $122^2 + 225 = 9^2 + 12^2$; $121^2 + 468 = 12^2 + 18^2$; $120^2 + 709 = 15^2 + 22^2$, $114^2 + 2113 = 32^2 + 33^2$, $111^2 + 2788 = 22^2 + 48^2$, $32^2 + 42^2$; $108^2 + 3445 = 9^2 + 58^2$, $14^2 + 57^2$, $41^2 + 42^2$.

15108: $122^2 + 224 = 2^5 \times 7$, 3 (mod 4), not the sum of two squares, $121^2 + 467$, 3 (mod 4), not the sum of two squares, $120^2 + 708$ = impossible; $119^2 + 947 = 3$ (mod 4), not the sum of two squares, $118^2 + 1184 = 20^2 + 28^2$, $116^2 + 1652$, not the sum of two squares, $112^2 + 2564 = 8^2 + 50^2$; after a lot of attempts: $86^2 + 76^2 + 44^2$.

15107: $122^2 + 223$, 3 (mod 4), not the sum of two squares; $121^2 + 466 = 5^2 + 21^2$, $119^2 + 946$ = not the sum of two squares, $117^2 + 1418 = 7^2 + 37^2$, $115^2 + 1882 = 19^2 + 39^2$; $105^2 + 4082 = 19^2 + 61^2$, $41^2 + 49^2$.

15106: $122^2 + 222 \div 2 = 111$, 3 (mod 4), not the sum of two squares, $121^2 + 465$ = not the sum of two squares, $120^2 + 706 = 9^2 + 25^2$; $119^2 + 945$, not the sum of two squares, $115^2 + 1881$, not the sum of two squares. The number is 4 (mod 9), so subtract an odd square 0 (mod 9) or 4 (mod 9), and then the possibilities are $101^2 + 69^2 + 12^2$; $99^2 + 72^2 + 11^2$ and $6241 + 8865 = 36^2 + 87^2$, $48^2 + 81^2$.

15105: $122^2 + 221 = 5^2 + 14^2$, $10^2 + 11^2$; $121^2 + 464 = 8^2 + 20^2$, $120^2 + 705$ = not the sum of two squares, $119^2 + 944 \div 16 = 59$, 3 (mod 4), not the sum of two squares, 118^2

$+ 1181 = 5^2 + 34^2$. The number is 3 (mod 9), so no squares of any number 0 (mod 3) are possible, $116^2 + 1649 = 7^2 + 40^2$, $25^2 + 32^2$.

15104: $122^2 + 220 \div 4 = 55$, 3 (mod 4), not the sum of two squares, $121^2 + 463 = 3$ (mod 4), not the sum of two squares, $120^2 + 704 = 26 \times 11$, 3 (mod 4), not the sum of two squares, $119^2 + 943$, 3 (mod 4), not the sum of two squares. Factorisation: $15104 = 28 \times 59 = 3$ (mod 4), so impossible. Not so nice, but possible is $80^2 + 80^2 + 48^2$.

15103: 7 (mod 8) and therefore it cannot be the sum of three squares.

15102: $122^2 + 218 = 7^2 + 13^2$, $121^2 + 461 = 10^2 + 19^2$, $120^2 + 702 \div 2 = 351$, 3 (mod 4), not the sum of two squares, $119^2 + 941 = 10^2 + 29^2$, $118^2 + 1178$, not the sum of two squares. The number is 6 (mod 8) and 0 (mod 9), so do not subtract a square 4 (mod 8); $115^2 + 1877 = 14^2 + 41^2$; $109^2 + 3221 = 14^2 + 55^2$, $107^2 + 3653 = 17^2 + 58^2$, $38^2 + 47^2$.

15101: $122^2 + 217 =$ not the sum of two squares, $121^2 + 460 \div 4 = 115 = 3$ (mod 4), not the sum of two squares, $120^2 + 701 = 5^2 + 26^2$, $101^2 + 4900 = 42^2 + 56^2$; $98^2 + 5497 = 23 \times 239$, each 3 (mod 4), not the sum of two squares. $15101 = 117^2 + 34^2 + 16^2$. $15101 = 93^2 + 76^2 + 26^2$.

15100: the basic number 151 is 7 (mod 8). You could try using only even squares, but there is no solution for the sum of three squares!

15099: 3 (mod 8), As this number is 3 (mod 8) the only possibilities lie in using three numbers each being 1 (mod 8) and no one 0 (mod 9). But not all of them give a good result. E.g. $119^2 + 938$ is no success. But here we come: $121^2 + 17^2 + 13^2$; $115^2 + 43^2 + 5^2$, $107^2 + 59^2 + 13^2$.

15098: $122^2 + 214 \div 2 = 207$, 3 (mod 4), not the sum of two squares; $121^2 + 457 = 4^2 + 21^2$; $120^2 + 698 = 13^2 + 23^2$; $119^2 + 937 = 19^2 + 24^2$; $118^2 + 1174 \div 2 = 587$, 3 (mod 4), not the sum of two squares; $117^2 + 1409 = 25^2 + 28^2$; $116^2 + 1642 = 11^2 + 39^2$.

15097: $122^2 + 213 \div 3 = 71$, 3 (mod 4), not the sum of two squares; $121^2 + 456$, not the sum of two squares, $120^2 + 11^2 + 24^2$, $16^2 + 21^2$; $119^2 + 936 = 6^2 + 30^2$; as 15097 is 4 (mod 9). The way to do this is either subtracting a square 0 (mod 9) or 4 (mod 9) and try; $117^2 + 1408 = 2^7 \times 11$, no solution; $120^2 + 11^2 + 24^2$, $16^2 + 21^2$; $106^2 + 3861$, not the sum of two squares; $101^2 + 4896 = 36^2 + 60^2$.

15096: $122^2 + 212 = 4^2 + 14^2$; $121^2 + 455 = 3$ (mod 4), not the sum of two squares, $120^2 + 696 \div 8 = 87$, 3 (mod 4), not the sum of two squares. As the number is 0 (mod

8) and 3 (mod 9), subtract only even squares which are not 0 (mod 9); $118^2 + 1172 = 4^2 + 34^2$; $116^2 + 1640 = 14^2 + 38^2$, $22^2 + 34^2$; $106^2 + 3860 = 4^2 + 62^2$, $34^2 + 52^2$.

15095: As this number is 7 (mod 8) there is no solution for the sum of three squares.

15094: $122^2 + 210 =$ not the sum of two squares, $120^2 + 694 \div 2 = 347$, 3 (mod 4), not the sum of two squares. As the number is 1 (mod 9); $118^2 + 1170 = 9^2 + 33^2$, $21^2 + 27^2$; $116^2 + 1638 \div 18 = 91$, 3 (mod 4), not the sum of two squares, $107^2 + 3645 = 27^2 + 54^2$; $98^2 + 5490 = 27^2 + 69^2$, $39^2 + 63^2$.

15093: $122^2 + 209 = 11 \times 19$, 3 (mod 4), not the sum of two squares, $121^2 + 452 = 14^2 + 16^2$; $120^2 + 693 \div 3 = 231$, 3 (mod 4), not the sum of two squares; $119^2 + 932 = 16^2 + 26^2$; $118^2 + 1169 = 7 \times 167$, 3 (mod 4), not the sum of two squares; as the number is 0 (mod 9): $116^2 + 1637 = 26^2 + 31^2$; $88^2 + 7349 = 25^2 + 82^2$.

15092: $122^2 + 208 = 8^2 + 12^2$, $121^2 + 451 = 3$ (mod 4), not the sum of two squares $120^2 + 692 = 4^2 + 26^2$. As the number is 0 (mod 4) and 8 (mod 9) subtract only an even square which is not 0 (mod 9); $118^2 + 1168 = 12^2 + 32^2$; $116^2 + 1636 = 6^2 + 40^2$.

15091: $122^2 + 207 = 3$ (mod 4), not the sum of two squares, $121^2 + 450 = 3^3 + 21^2$; $120^2 + 691 = 3$ (mod 4), not the sum of two squares; $117^2 + 1402 = 21^2 + 31^2$; $8649 + 6442 = 41^2 + 69^2$.

15090: $122^2 + 206 \div 2 = 103$, 3 (mod 4), not the sum of two squares; $121^2 + 449 = 7^2 + 20^2$, $120^2 + 690$, not the sum of two squares; $119^2 + 929 = 20^2 + 23^2$; the number is 6 (mod 9) so do not subtract numbers 0 (mod 9); $103^2 + 4481 = 16^2 + 65^2$; $101^2 + 4889 = 20^2 + 67^2$.

15089: $122^2 + 205 = 3^2 + 14^2$, $6^2 + 13^2$, $121^2 + 448 = 64 \times 7$, so impossible, $120^2 + 689 = 8^2 + 25^2$, $17^2 + 20^2$, $119^2 + 928 = 12^2 + 28^2$; $117^2 + 1400$, not the sum of two squares, $115^2 + 1864 = 10^2 + 42^2$; $113^2 + 2329 = 5^2 + 48^2$, $27^2 + 40^2$; $111^2 + 2768 = 8^2 + 52^2$; $109^2 + 3208 = 38^2 + 42^2$.

15088: This number is divisible by 16, 943 times. As $943 = 7$ (mod 8) there is no solution for the sum of three squares.

15087: the number is 7 (mod 8), so no solution for the sum of 3 squares.

15086: $122^2 + 202 = 9^2 + 11^2$; $120^2 + 686 \div 2 = 343 = 7$ (mod 8), not the sum of two squares; $119^2 + 925 = 5^2 + 30^2$, $14^2 + 27^2$, $21^2 + 22^2$; $118^2 + 1162$, not the sum of two squares, $115^2 + 1861 = 30^2 + 31^2$; $109^2 + 3205 = 17^2 + 54^2$, $107^2 + 3637 = 39^2 + 46^2$. With numbers such as this one, always work with two odd squares.

15085: $122^2 + 201 =$ not the sum of two squares, $121^2 + 444 \div 4 = 111 = 3 \pmod 4$, not the sum of two squares, $120^2 + 685 = 3^2 + 26^2, 18^2 + 19^2, 119^2 + 924 \div 4 = 231 = 3 \pmod 4$, not the sum of two squares, $18^2 + 1161 =$ not the sum of two squares; $117^2 + 1396 = 10^2 + 36^2$; $116^2 + 1629 = 27^2 + 30^2, 115^2 + 3860 = 4^2 + 62^2, 34^2 + 5^2$.

15084: $122^2 + 200 = 2^2 + 14^2, 120^2 + 684 \div 4 = 171, 3 \pmod 4$, not the sum of two squares. The number is $0 \pmod 4$ and $0 \pmod 9$, so subtract an even square; $118^2 + 1160 = 2^2 + 34^2, 22^2 + 26^2, 116^2 + 1628 \div 4 = 407, 3 \pmod 4$, not the sum of two squares; $114^2 + 2088 = 18^2 + 42^2$; $110^2 + 2984 = 22^2 + 50^2$; $106^2 + 3848 = 2^2 + 62^2, 22^2 + 58^2$.

8.9.4 A surprise!!

After having found something very special in the sum of four squares, the question came up "Is something comparable happening in the sums of three squares?" And indeed it is. In the table you see four times the sum of 324 and the sum of the squares is 50,000, in the same way we see four times the sum of 348 and the sum of the squares 50,000 too.

220	+	32	+	24	50000	276
208	+	60	+	56	50000	324
200	+	96	+	28	50000	324
188	+	120	+	16	50000	324
160	+	156	+	8	50000	324
200	+	80	+	60	50000	340
196	+	80	+	72	50000	348
192	+	100	+	56	50000	348
176	+	132	+	40	50000	348
160	+	152	+	36	50000	348
184	+	112	+	60	50000	356
176	+	120	+	68	50000	364
168	+	124	+	80	50000	372
160	+	120	+	100	50000	380

What we see here is from left to right three columns with the basic numbers, then the sum of their squares (50,000), and in the most right column the sum of the basic numbers. And what is so nice to see? We see that four times the sum of the different basic numbers is equal and that the sum of their squares unexpectedly also is equal.

And we see that this nice situation happens with two sums; the sum of 324 and the sum of 348.

We did not see this if the sum of two squares, but in this situation the number of combinations is considerably lower. If we take 100,000 as the upper limit and derived possibilities are not taken in account, there are never more than eight combinations possible.

8.10 Questions

Find three squares ending in 1236.

Find four squares ending in 8169.

Find four squares ending in 37184.

The following numbers are each the sum of three different squares: 1837; 2321; 2867. Find them.

8.11 Solutions

$106^2 = 11236$. This is already the first one! As we work here in the four digit area, we take a quarter of 10,000, i.e. 2500. Next we take 2500 − 106 and get 2394 and 2500 + 106 to get 2606. Make the multiplications and check the results!

We know that $13^2 = 169$. Based on $(a + b)^2$ we calculate that 1013^2 ends on 6169, and that each additional thousand results in an increase of 6 in the thousands of the result. So our first candidate is the number 3013. Again working in the four digit area we can take 5000 − 3013 = 1987. This number squared results in 394 8169. $63^2 = 3969$ and based on $(a + b)^2$ we know the increase in value is 2600 per 100 in the basic number. We are at 3969 and have to go to 8169; the difference being 4200. Now we divide 42 ÷ 26 = 17, and calculate the square of 1763, which is (310) 8169. We now have 1763, 1987 and 3013. The fourth number is 10,000 − 8013 = 1987.

72^2: 5184; for 37,184 we have to increase the thousands by 32,000. Based on the $(a + b)^2$ we calculate that for 72 the thousands increase with by 44. Now: 32 ÷ 144 = 3, 28, 53 or 78. We start with the smallest one: 3072. As we work in the five digit area we take 100,000 ÷ 4 = 25,000, then subtract 25,000 − 3,072 = $21,928^2$ = (4808) 37184. We continue with 25,000 + 3072 = 28,072 finish with 50,000 + 3072 = 46,928. Check the results!!

1837: Next lower square (NLS) is 1764: a lucky strike as $1837 - 1764$ (42^2) = 73 = $8^2 + 3^2$. $1837 - 41^2 = 1681 = 156$ which is 3 (mod 9) so no option; Likewise with $-39^2 = 316$, no go; $-38^2 = 393$, no go; $-37^2 = 468 = 18^2 + 12^2$; $-36^2 = 541 = 21^2 + 10^2$.

2321: NLS = 47^2 = 2209, difference 112, no go; $-46^2 = 205$, which can be written as $14^2 + 3^2$ and $13^2 + 6^2$. 2321. Then $-45^2 = 296 = 14^2 + 10^2$; $-44^2 = 385$, no go; $-43^2 = 472$, no go; $-42^2 = 557 = 19^2 + 14^2$. $-41^2 = 640 = 24^2 + 8^2$.

2867: NLS = $53^2 = 2809$. $2867 - 2809 = 58 = 7^2 + 3^2$. $2867 - 52^2 = 183 = 7$ (mod 8), no go. $2867 - 51^2 = 286 ÷ 2 = 143 = 7$ (mod 8), no go, $2867 - 50^2 = 387 = 3$ (mod 8) no go, $-49^2 = 486$, $-48^2 = 583 = 7$ (mod 8) no go, $-47^2 = 678 = 6$ (mod 8) no go $-46^2 =$

751 = 7 (mod 8) no go; -45² = 842 = 29² + 1². Other numbers which drop off: 44², 43, 42². Positive is 2867 – 41² = 1186 = 31² + 15².

8.12 Sums of four squares

In the book "The great mental calculators" we read examples of calculations prodigies who are able to write a number of five or even more figures in a few seconds as the sum of four squares.

As written before we take the same numbers as with the sum of three squares. Now we have still more possibilities but we will not describe all the possibilities. We will see that where the sum of three squares appeared to be impossible, these numbers can yet be written as the sum of four squares.

Some data:

About Gottfried Rückle (1879 – 1929), we know that he succeeded in writing the following numbers as the sum of four squares:

11,339 = 105² + 15² + 8² + 5² in 56 seconds

18,111 = 134² + 11² + 5² + 3² in 26.5 seconds
       = 134² + 9² + 7² + 5² in 63.5 seconds

53,116 = 230² + 14² + 4² + 2² in 51 seconds
       = 230² + 12² + 6² + 6² immediately after

About Wim Klein (1913 – 1986) we know that he could think of 12 different solutions for the number 5359 in about one minute. Somewhat later he did ten different numbers in 1 minute 14 seconds.

Later on he wanted to set a record; ten different numbers within a minute. This never happened; in August 1986 he was brought to death by violence, the murderer has never been found.

We will now work with the same numbers as with the sum of three squares.

Here also we start with the nearest lower square. But if a number 7 mod 8 remains, we can stop immediately, as we already know that such a number cannot be the sum of three squares.

8.13 The numbers

15120: 122² + 14² + 6² + 2²; 121² + 21² + 3² + 3²; 120² + 20² + 16² + 8²; 118² + 30² +

$14^2 + 10^2$

15119: $122^2 + 15^2 + 3^2 + 1^2$; $121^2 + 17^2 + 10^2 + 7^2$; $119^2 + 27^2 + 15^2 + 2^2$

15118: $122^2 + 12^2 + 9^2 + 3^2$; $121^2 + 19^2 + 10^2 + 4^2$; $120^2 + 718$, no go; $119^2 + 19^2 + 14^2 + 20^2$; $118^2 + 25^2 + 13^2 + 20^2$; $117^2 + 24^2 + 18^2 + 23^2$; $116^2 + 37^2 + 17^2 + 2^2$

15117: $122^2 + 14^2 + 6^2 + 1^2$; $121^2 + 476$ no solution; $120^2 + 19^2 + 16^2 + 10^2$; $119^2 + 956$ no solution; $118^2 + 33^2 + 10^2 + 2^2$; $117^2 + 34^2 + 16^2 + 4^2$; $116^2 + 35^2 + 6^2 + 20^2$; $115^2 + 36^2 + 20^2 + 14^2$

15116: $122^2 + 232, 6^2 + 14^2$; $121^2 + 21^2 + 5^2 + 3^2$; $120^2 + 22^2 + 14^2 + 6^2$; $119^2 + 27^2 + 15^2 + 1$; $118^2 + 30^2 + 16^2 + 6^2$

15115: $122^2 + 231$ no solution as 7 (mod 8); $121^2 + 20^2 + 7^2 + 5^2$; $120^2 + 25^2 + 9^2 + 3^2$; $119^2 + 23^2 + 5^2 + 20^2$; $118^2 + 1191$ no solution as 7 (mod 8); $117^2 + 36 + 9^2 + 7^2$

15114: $122^2 + 15^2 + 2^2 + 1^2$; $121^2 + 18^2 + 7^2 + 10^2$; $120^2 + 25^2 + 8^2 + 5^2$; $119^2 + 28^2 + 12^2 + 5^2$

15113: $122^2 + 12^2 + 9^2 + 2^2$; $121^2 + 30^2 + 6^2 + 4^2$; $120^2 + 18^2 + 17^2 + 10^2$; $119^2 + 18^2 + 12^2 + 2^2$

15112: $122^2 + 14^2 + 4^2 + 4^2$; of $122^2 + 10^2 + 8^2 + 8^2$; $121^2 + 21^2 + 3^2 + 1^2$; $120^2 + 24^2 + 10^2 + 6^2$; $119^2 + 951$, no solution as 7 (mod 8); $118^2 + 32^2 + 10^2 + 8^2$; $117^2 + 1423$ no solution as 7 (mod 8); $116^2 + 36^2 + 18^2 + 6^2$

15111: $122^2 + 13^2 + 7^2 + 3^2$; $121^2 + 16^2 + 14^2 + 2^2$; $120^2 + 711$ no solution as 7 (mod 8); $119^2 + 19^2 + 3^2 + 10^2$

15110: $122^2 + 12^2 + 9^2 + 1^2$; $121^2 + 15^2 + 12^2 + 10^2$; $120^2 + 23^2 + 9^2 + 10^2$; $119^2 + 24^2 + 18^2 + 7^2$

15109: $122^2 + 12^2 + 9^2 + 2^2$; $121^2 + 20^2 + 8^2 + 2^2$; $120^2 + 22^2 + 12^2 + 9^2$; $119^2 + 28^2 + 8^2 + 10^2$

15108: $122^2 + 12^2 + 8^2 + 4^2$; $121^2 + 21^2 + 5^2 + 1^2$; $120^2 + 26^2 + 4^2 + 4^2$; $119^2 + 29^2 + 9^2 + 5^2$; $118^2 + 28^2 + 16^2 + 12^2$

15107: $122^2 + 223$, 7 (mod 8) so not the sum of three squares, $121^2 + 21^2 + 4^2 + 3^2$; $120^2 + 25^2 + 9^2 + 1^2$; $119^2 + 27^2 + 14^2 + 5^2$; $118^2 + 1183$ which is 7 (mod 8), so not the sum of three squares

15106: $122^2 + 14^2 + 5^2 + 1^2$; $121^2 + 19^2 + 10^2 + 2^2$; $120^2 + 24^2 + 7^2 + 9^2$; $119^2 + 30^2 + 6^2 + 3^2$

15105: $122^2 + 14^2 + 4^2 + 3^2$; $121^2 + 16^2 + 12^2 + 8^2$; $120^2 + 26^2 + 5^2 + 2^2$; $119^2 + 28^2 + 12^2 + 4^2$

15104: $122^2 + 220 \div 4 = 55$, 7 (mod 8) so not the sum of three squares; $121^2 + 463$, 7 (mod 8), so not the sum of three squares; $120^2 + 24^2 + 8^2 + 8^2$; $119^2 + 943$, 7 (mod 8), so not the sum of three squares; $118^2 + 1180 \div 4 = 295$, 7 (mod 8), so not the sum of three squares; $117^2 + 1415$, 7 (mod 8), so not the sum of three squares; the number 28×59 leaves no solution

15103: $122^2 + 13^2 + 7^2 + 1^2$; $121^2 + 19^2 + 10^2 + 1^2$; $120^2 + 703$, 7 (mod 8), so not the sum of three squares; $119^2 + 29^2 + 10^2 + 1^2$; $118^2 + 29^2 + 17^2 + 7^2$

15102: $122^2 + 7^2 + 12^2 + 5^2$, $121^2 + 19^2 + 8^2 + 6^2$; $120^2 + 26^2 + 5^2 + 1^2$; $119^2 + 21^2 + 20^2 + 10^2$

15101: $122^2 + 9^2 + 6^2 + 10^2$; $121^2 + 18^2 + 10^2 + 4^2$; $120^2 + 24^2 + 10^2 + 5^2$; $119^2 + 30^2 + 6^2 + 2^2$

15100: $122^2 + 10^2 + 10^2 + 4^2$; $121^2 + 17^2 + 13^2 + 1^2$; $120^2 + 700, \div 4 = 175 = 7$ (mod 8), so not the sum of three squares; $119^2 + 25^2 + 17^2 + 5^2$

15099: $122^2 + 215 = 7$ (mod 8), so not the sum of three squares; $121^2 + 17^2 + 12^2 + 5^2$; $120^2 + 25^2 + 7^2 + 5^2$; $119^2 + 23^2 + 20^2 + 3^2$

15098: $122^2 + 13^2 + 6^2 + 3^2$; $121^2 + 15^2 \, 14^2 + 6^2$; $120^2 + 19^2 + 11^2 + 4^2$; $119^2 + 30^2 + 6^2 + 1^2$

15097: $122^2 + 14^2 + 4^2 + 1^2$; $121^2 + 16^2 + 14^2 + 2^2$; $120^2 + 25^2 + 6^2 + 6^2$. $697 = 4$ (mod 9), which means one of the squares has to be 4 (mod 9). There are many "doubles" also: $18^2 + 18^2 + 7^2$; $119^2 + 24^2 + 18^2 + 6^2$

15096: $122^2 + 12^2 + 8^2 + 2^2$; $121^2 + 16^2 + 10^2 + 10^2$ of $14^2 + 14^2 + 8^2$; $120^2 + 20^2 + 10^2 + 14^2$; $119^2 + 935 = 7$ (mod 8), not the sum of three squares; $118^2 + 30^2 + 16^2 + 4^2$; $117^2 + 1407 = 7$ (mod 8), not the sum of three squares; $116^2 + 40^2 + 6^2 + 2^2$

15095: $122^2 + 11^2 + 9^2 + 3^2$; $121^2 + 24^2 + 7^2 + 3^2$; $120^2 + 695 = 7$ (mod 8), not the sum of three squares; $119^2 + 30^2 + 5^2 + 3^2$

15094: $122^2 + 13^2 + 5^2 + 4^2$; $121^2 + 20^2 + 7^2 + 2^2$; $120^2 + 18^2 + 17^2 + 9^2$; $119^2 + 28^2 + 10^2 + 7^2$

15093: $122^2 + 14^2 + 3^2 + 2^2$; $121^2 + 20^2 + 6^2 + 4^2$; $120^2 + 23^2 + 10^2 + 8^2$; $119^2 + 24^2 + 16^2 + 10^2$

15092: $122^2 + 12^2 + 8^2 + 0^2$ (not so very nice, this 0); $121^2 + 21^2 + 3^2 + 1^2$; $120^2 + 24^2 + 10^2 + 4^2$; $119^2 + 29^2 + 9^2 + 3^2$

15091: $122^2 + 207 = 7$ (mod 8), not the sum of three squares; $121^2 + 21^2 + 3^2 + 0^2$ of $15^2 + 15^2 + 0^2$, again a zero; $120^2 + 21^2 + 15^2 + 5^2$; $119^2 + 29^2 + 8^2 + 5^2$

15090: $122^2 + 10^2 + 9^2 + 5^2$; $121^2 + 16^2 + 12^2 + 7^2$; $120^2 + 25^2 + 8^2 + 1^2$; $119^2 + 23^2 + 16^2 + 12^2$

15089: $122^2 + 12^2 + 6^2 + 5^2$; $121 + 448 = 26 \times 7$, 7 (mod 8), not the sum of three squares; $120^2 + 24^2 + 8^2 + 7^2$; $119^2 + 928 = 25 \times 29$, not the sum of three squares; $118^2 + 30^2 + 16^2 + 3^2$

15088: $122^2 + 14^2 + 2^2 + 2^2$; $121^2 + 447 = 7$ (mod 8), not the sum of three squares; $120^2 + 20^2 + 12^2 + 12^2$; $119^2 + 927 = 7$ (mod 8), not the sum of three squares; $118^2 + 34^2 + 2^2 + 2^2$; $117^2 + 1399 = 7$ (mod 8), not the sum of three squares; $116^2 + 40^2 + 4^2 + 4^2$

15087: $122^2 + 11^2 + 9^2 + 1^2$; $121^2 + 15^2 + 11^2 + 10^2$; $120^2 + 687 = 7$ (mod 8), not the sum of three squares; $119^2 + 926 \div 2 = 463 = 7$ (mod 8), not the sum of three squares; $118^2 + 33^2 + 7^2 + 5^2$

15086: $122^2 + 12^2 + 7^2 + 3^2$; $121^2 + 25^2 + 4^2 + 2^2$; $120^2 + 25^2 + 6^2 + 5^2$; $119^2 + 24^2 + 18^2 + 5^2$

15085: $122^2 + 14^2 + 2^2 + 1^2$; $121^2 + 444 \times 4 = 111, = 7$ (mod 8), not the sum of three squares; $120^2 + 24^2 + 10^2 + 3^2$; $119^2 + 924 \div 4 = 231 = 7$ (mod 8), not the sum of three squares; $118^2 + 30^2 + 15^2 + 6^2$

15084: $122^2 + 10^2 + 8^2 + 6^2$; $121^2 + 21^2 + 1^2 + 1^2$; $120^2 + 22^2 + 14^2 + 2^2$; $119^2 + 29^2 + 9^2 + 1^2$; $118^2 + 30^2 + 16^2 + 2^2$

You see: We stayed close to the question numbers, so none of the enumerations is complete. To get an impression about how many combinations were possible, I took the number 49,999 and investigated how many times this number can be written as the sum of four squares, whereby if possible the double squares and zeros are kept out of consideration.

8.14 A small inquiry: 49,999 as the sum of four squares.

To get informed about the sum of four squares I took the number 49,999 to examine how many times this number can be written as the sum of four squares. The possibility of 0 was ignored, however sometimes a square is taken twice: $49,999 = 221^2 + 34^2 + 1^2 + 1^2$.

223	+	15	+	6	+	3	49999
223	+	13	+	10	+	1	49999
223	+	11	+	10	+	7	49999
222	+	25	+	9	+	3	49999
222	+	21	+	15	+	7	49999
221	+	34	+	1	+	1	49999
221	+	31	+	14	+	1	49999
221	+	29	+	14	+	11	49999
221	+	25	+	23	+	2	49999
221	+	25	+	22	+	7	49999

In total I counted 1086 combinations, many more than I presumed – my estimation was ± 600 – the work took me about eleven hours.

I spare you the grief of the complete table. It is a matter of counting-out rhyme from bigger numbers to smaller ones; starting with the nearest lower square and then counting down.

We start with 223^2 and subtract this from 49999, result 270. Now it is the matter to write this as the sum of three squares and we find:

- $15^2 + 6^2 + 3^2$
- $14^2 + 7^2 + 5^2$
- $13^2 + 10^2 + 1^2$
- $11^2 + 10^2 + 7^2$

So with 223 we have four possibilities, and go on to 222^2. $49,999 - 222^2 = 715$ which we can write as:

- $25^2 + 9^2 + 3^2$
- $21^2 + 15^2 + 7^2$

Gradually we get a table like the one below:

All these data were put in an Excel file and for checking there was a formula written in which the squares were added. Next the double solutions were cast out, quite a lot of work when there was a rather big number of combinations, for example: $211^2 + 73^2 + 7^2 + 10^2$ and $211^2 + 10^2 + 7^2 + 73^2$ are if included fully equal.

The following things struck me the most:
- The big number of possible combinations.
- The possibility of no solution to be found was considered: if 49999 minus the square would result in a number 7 mod 8 there would be no solution possible as a number 7 mod 8 cannot be written as the sum of three squares. There should be about thirty of these numbers; this is confirmed by the facts.
- It was a very interesting challenge.

- As the nearest lower square gets smaller, the number of possibilities increases considerably: with 175 + three other squares there are no fewer than 29 possibilities, in the 200 series a lot fewer!.
- Only one time was a square is counted three times: $182^2 + 3 \times 75^2 = 49{,}999$
- 49.999 is 4 mod 9 and 15 mod 16. All the combinations have to obey the following conditions: $1 \times 9 + 1 \times 4 + 1 \times 1 + 1 \times 1$ mod 16, for example as the squares of 205, 58, 49 and 47.

This is illustrated by this table:

Basic number	Mod 9	Mod 16
205^2	4	9
58^2	7	4
49^2	7	1
47^2	4	1
Sum	4	15

The big table below informs you about the possibilities depending on the nearest lower square:

$49.999 - x^2$	Number of combi's	$49.999 - x^2$	Number of combi's
223	4	170	16
222	2	169	29
221	7	168	0
220	0	167	24
219	5	166	10
218	9	165	23
217	10	164	0
216	0	163	20
215	12	162	14
214	12	161	21
213	9	160	0
212	0	159	12
211	10	158	24
210	10	157	37
209	12	156	0
208	0	155	34
207	15	154	17
206	13	153	12
205	24	152	0
204	0	151	27
203	16	150	16
202	10	149	13
201	10	148	0
200	0	147	13
199	13	146	9
198	6	145	19
197	16	144	0
196	0	143	17
195	17	142	7
194	10	141	12
193	24	140	0
192	0	139	8
191	18	138	6
190	15	137	19
189	18	136	0
188	0	135	9
187	21	134	6
186	9	133	7
185	28	132	0
184	0	131	4
183	14	130	5
182	26	129	4
181	16	128	0
180	0	127	4
179	22	126	1
178	20	125	4
177	15	124	0
176	0	123	1
175	29	122	1
174	8	121	1
173	25	120	0
172	0	116	0
171	18	115	2
Subtotal	578	Total	1086

Finally: 49,999 is a prime number 3 mod 4 and 7 mod 8 and therefore it can neither be written as the sum of two squares nor as the sum of three different squares because of 7 mod 8. You'll find this confirmed in the table: behind every basic number 0 mod 4 you'll see that the number of combinations is 0.

8.15 Surprise, Surprise!

The numbers for which the squares give the sum 49,999 were put in a table, which was quite a job. It was evident that between them were examples of doubles: $173^2 + 107^2 + 70^2 + 61^2$ and $173^2 + 70^2 + 107^2 + 61^2$ are, if included, fully equal. These had to be cast out. This idea came up: count the sum of the basic numbers and sort by value. This resulted in the below standing table:

173	+	107	+	70	+	61	49999	411
170	+	91	+	103	+	47	49999	411
167	+	113	+	85	+	46	49999	411
167	+	43	+	106	+	95	49999	411
161	+	115	+	97	+	38	49999	411
161	+	103	+	110	+	37	49999	411
158	+	127	+	85	+	41	49999	411
158	+	95	+	121	+	37	49999	411
158	+	85	+	127	+	41	49999	411
157	+	125	+	91	+	38	49999	411
157	+	101	+	118	+	35	49999	411

And look: all these combinations are different and yet the sums of their squares are equal!

8.16 Another small examination: 50,000 as sum of four squares

Getting still more curious by examining 49,999 the question came up: How will it be with 50,000? Big question: How many or how few combinations are possible as the sum of four squares? First thought; there should be fewer possibilities because the sum of 0 mod 16 can only be found by four times 0 mod 16 or four times 4 mod 16 and by consequence all the odd numbers are to cast out.

There are in total 260 combinations, do you see the big difference with 49,999? Well, 50,000 is 5 mod 9 which can be obtained as follows: $1 \times 4 + 1 \times 1 + 2 \times 0$ or $2 \times 7 + 2 \times 0$ mod 9.

For example the sum of the squares 160 + 136 + 60 + 48 or 130 + 162 + 66 + 50.

For elucidation, this table:

Basic number	Mod 9	Mod 16	Basic number	Mod 9	Mod 16
160²	4	0	130²	7	4
136²	1	0	162²	0	4
60²	0	0	66²	0	4
48²	0	0	50²	7	4
Sum	5	0	Sum	5	0

The number 50,000 can be written three times as the sum of two different squares, and sixteen times the sum of three squares, and as you already know 260 times the sum of four squares.

As the smallest sum of the basic numbers I found 247: 223 + 15 + 6 + 3, as the biggest sum 447: 115 + 113 + 113 + 106.

This difference is understandable: Nearest lower square 223^2 = 49,729 so we need only 270 to get 49,999. Again nearest lower square 16^2 = 256, we are almost there.

The biggest sum: 49,999 ÷ 4 = 12,499, close to 112^2 = 12,544. And 4 × 112 = 448.

Is that your experience too? You found something which works according to some rule or formula. And then look if there is – more or less hidden – some exception. And it appears there is no exception, so that the method followed is correct.

9 Integer square roots

The quantity of work to do depends on the knowledge of the squares of the mental calculator. We'll work in 3 stages. The knowledge of the squares 1–10 is assumed to be there; without even that nothing will be possible.

```
          4   9   1
          4   9   1
      ─────────────
              1        (1)  1 × 1
          (1)8          (2)  1 × 9 + 9 × 1 + 1
        (9)0            (3)  1 × 4 + 1 × 4 + 9 × 9 + 1
      (8)1              (4)  9 × 4 + 4 × 9 + 9
   24                   (5)  4 × 4 + 8
   ─────────────────
   24   1   0   8   1
```

$$\sqrt{241081}$$

This is an integer root.

- We start with 24. N(earest) L(ower) S(quare) 16, so the answer so far is 4
- 24 − 16 = 8 ÷ 2 = 4 × 10 = 40 + append half the next digit (1) ÷ 2 = 40 + 0.5 = 40.5. If we calculate here as the result 10 r 0.5 we get into trouble later on. As we have to multiply and after that to subtract, there could easily be a negative result, with which we cannot continue. Therefore we round down and write 40.5 ÷ 4 = 9 r 4.5. Answer so far 49
- 4.5 × 10 = 45 + half next digit 0 ÷ 2 = 45 − (9^2 ÷ 2) = 4.5 ÷ 4 = 1 r 0.5. Answer so far 491
- 0.5 × 10 = 5 + append half next digit = 8 ÷ 2 = 4, sum = 9 − 9 × 1 = 0 ÷ 4 = 0 r 0. Answer so far 491.0
- 0 × 10 = 0 + append half next digit = 1 ÷ 2 = 0.5 − (0 × 9) − 1^2 ÷ 2 = 0.0 ÷ 4 = 0 r 0, final answer 491.00.

Study this way of working which is derived from the cross method in an inverse way.

We do this step by step:

- Subtract the NLS from the digits we start with, here 24 – 16, first digit of the answer, here 4
- We divide the remainder 24 – 16 = 8 by 2 and multiply this number × 10
- Result of multiplication + append the half of the next digit, add this
- Addition divided by the first digit of the answer gives the next digit of the answer, here 49
- Remainder of the division × 10 + append half next digit – half square of next digit, here $9^2 \div 2$
- Divide this – 45 – $9^2 \div 2$ – by (here) 4, here 1 r 0.5
- Remainder of the division × 10 + half next digit – 9 × 1 and divide by 4. 9 – 9 = 0 ÷ 4 = 0 r 0 Answer 491.0
- In fact you can stop here as the question is an integer square root, the comma/decimal point + decimal digits are not required.

After having found the first digit(s) of the answer it is a matter of:

- Remainder × 10 + half next digit
- Then, multiply the digits after the divider as follows:
- In the case of an odd number of digits after the divider take the odd ones squared and divide by 2
- In the case of an even number of digits after the divider multiply according to the cross method and subtract the result of the remainder × 10

$$\sqrt{241081}$$

We do now the same question with a bigger knowledge of the squares:

- 2410 – 49^2 (2401) = 9, answer so far 49
- Half of the remainder × 10 = 4.5 × 10 = 45, only the first remainder should be divided by 2
- 45 + add half next digit (8 ÷ 2) = 4, 45 + 4 = 49 and 49 ÷ 49 = 1 r 0, answer so far 491
- Remainder × 10 = 0 + append half next digit = 0.5 – $1^2 \div 2$ = 0 ÷ 49 = 0, final answer 491.0

For those who know the squares up to 1,000, here is a bigger example:

$$\sqrt{1073807361}$$

- 107,380 – (NLS 327^2 = 106,929) = 451, answer so far 327
- 451 ÷ 2 = 225.5 ÷ 32 = 6 r 33.5, answer so far 3276. Here again we round down, because the answer 7 r 1,5 would leave us in trouble, as later on a subtraction would leave a negative value.
- 33.5 × 10 = 335 + append half next digit 7 ÷ 2 = 338.5 – 6 × 7 = 296.5
- 296.5 ÷ 32 = 9 r 8.5, answer so far 32,769
- 8.5 × 10 = 85 + append half next digit 3 ÷ 2 = 86.5 – 7 × 9 – 6^2 ÷ 2 = 5.5
- 5.5 ÷ 32 = 0 r 5.5, answer so far 32,769.0
- 5.5 × 10 = 55 + append half next digit 6 ÷ 2 = 3, sum = 58 – 0 × 7 (for all clarity) – 6 × 9 = 4 ÷ 32 = 0 r 4, answer so far 32,769.00
- 4 × 10 = 40 + append half next digit 1 ÷ 2 = 40.5 – 0 × 7 – 0 × 6 – 9^2 ÷ 2 = 40.5 – 40.5 = 0 ÷32 = 0 r 0, answer 32769.000
- Again: As the question was an integer square root, the zeros are not required; one can stop at 32,769.

9.1 Decimal square roots

In this chapter we see the extent to which knowledge of the squares is helpful for this operation.
We'll start with a little knowledge.
Anyhow: The estimation should always be made in such a way that the result of the subtraction is never negative, and the result of the division may not exceed 10.

$$\sqrt{241861}$$

This number has six digits, so the answer will have 3 + the numbers of decimals we want to calculate. We start with 24 and subtract N(earest) L(ower) S(quare) 16, so the answer so far is 4

- 24 – 16 = 8 ÷ 2 = 4 × 10 = 40 + append half next digit 1 ÷ 2 = 0.5 = 40.5 ÷ 4 = 9 r 4.5. Answer so far 49.
- 4.5 × 10 = 45 + append half next digit 8 divided by 2, together 49 – (9^2 ÷ 2) 40.5 = 8.5 ÷ 4 = 1 r 4.5. Answer so far 491.
- 4.5 × 10 = 45 + append half next digit (6 ÷ 2) 3 = 48 – 9 × 1 = 39 ÷ 4 = 7 r 11. Answer so far 491.7
- 11 × 10 = 110 + append half next digit (1 ÷ 2) 0.5 = 110.5 – 1^2 ÷ 2 – 7 × 9 = 47 ÷ 4 = 9 r 11. Answer so far 491.79. As 1 is the last figure of the question number, there is nothing more to append.
- 11 × 10 = 110 – 9 × 9 – 7 × 1 = 22 ÷ 4 = 3 r 10. Answer so far 491.793

- $10 \times 10 = 100 - 9 \times 3 - 1 \times 9 - (7^2 \div 2) = 39.5 \div 4 = 6$ r 15.5. Answer so far 491.7936.
- $15.5 \times 10 = 155 - 6 \times 9 - 3 \times 1 - 9 \times 7 = 35 \div 4 = 5$ r 15. Answer so far 491.79365
- $15 \times 10 = 150 - 5 \times 1 - 6 \times 1 - 3 \times 7 - (9^2 \div 2) = 37.5 \div 4 = 5$ r 17.5. Answer so far 491.793655.

As in this type of questions generally five decimals are required, the answer is rounded as 491.79366. We now see that working in this way requires a lot of estimation: the more decimals are obtained, the bigger the subtractions become. To reduce this difficulty a greater knowledge of the squares is not only desirable but almost a necessity.

This way of working is the "inverse cross method". After you have calculated the first figure(s), there are two possibilities:

For an even number of decimals: Multiply 'mirror-wise' these decimals and subtract the result of the remainder × 10, and then divide

For an odd number of decimals: Multiply mirror-wise these decimals, calculate half the square of the remaining figure and subtract the result of the remainder × 10 and divide.

With a bigger knowledge of the squares, now up to 100, we do this:

$$\sqrt{241861}$$

NLS below 2418 = 2418 = 2401 = 49^2. Answer so far 49.

$2418 - 2401 = 17 \div 2 = 8.5 \times 10 = 85 + 6 \div 2 = 88 \div 49 = 1$ r 39. Answer so far 491

$39 \times 10 = 390$ and append $1 \div 2 = 390.5$ minus $(1^2 \div 2)$ $0.5 = 390 \div 49 = 7$ r 47. Answer so far 491.7.

$47 \times 10 + 0 \div 2 - 1 \times 7 = 463 \div 49 = 9$ r 22. Answer so far 491.79.

$220 - 1 \times 9 - 7^2 \div 2 = 186.5 : 49 = 3$ r 39.5. Answer so far 491.793.

$39.5 \times 10 = 395 - 3 \times 1 - 7 \times 9 = 329 \div 49 = 6$ r 35. Answer so far 491.7936.

$35 \times 10 = 350 - 1 \times 6 - 7 \times 3 - 9^2 \div 2 = 282.5 \div 49 = 5$ r 37.5. Answer so far 491.79365.

For rounding to 5 decimal places we look for another decimal. $37.5 \times 10 = 375 - 5 \times 1 - 7 \times 6 - 3 \times 9 = 301 \div 49 = 6$ r 7. Answer so far 491.793656, Rounded to 5 d.p., this is 491.79366.

With yet more knowledge of the squares, up to 1000, the first step is made still faster.

$$\sqrt{241861}$$

$491^2 = 241081$, answer so far 491.

$241861 - 241081 = 780 \div 2 = 390 \div 49 = 7$ r 47. Answer so far 491.7

$47 \times 10 = 470 - 1 \times 7 = 463 \div 49 = 9$ r 22. Answer so far 491.79

$22 \times 10 = 220 - 1 \times 9 - 7^2 \div 2 = 186.5 \div 49 = 3$ r 39.5. Answer so far 491.793

$39.5 \times 10 = 395 - 1 \times 3 - 7 \times 9 = 329 \div 49 = 6$ r 35. Answer so far 491.7936.

$35 \times 10 = 350 - 1 \times 6 - 3 \times 7 - 9^2 \div 2 = 282.5 \div 49 = 5$ r 37.5. Answer so far 491.79365

$37.5 \times 10 = 375 - 5 \times 1 - 6 \times 7 - 3 \times 9 = 281 \div 49 = 5$ r 36. Answer so far 491.79366 when rounded.

There is still another technique; it is up to the reader if this is preferable. The steps are bigger, because there are multiplications by 100, but now you get two decimals at a time. It is questionable if this method works faster. Anyhow a profound knowledge of the multiplication table up to 100 is required.

9.2 Multiplying by 100

$$\sqrt{241861}$$

2418, NLS = 2401, answer so far 49.

$2418 - 2401 = 17 \times 100 +$ append $61 = 1761 \div 2 = 880.5 \div 4\,9 = 17$ r 47.5. Answer so far 491.7.

$47.5 \times 100 = 4750 - (17^2 \div 2)\ 144.5 = 4605.5 \div 49 = 93$ r 48.5. Answer so far 491.793.

$48.5 \times 100 = 4850 - (17 \times 93)\ 1581 = 3269 \div 49 = 66$ r 35. Answer so far 491.79366.

The remainder is rather big. Nevertheless, if we should round down, this would be the result: $35 \times 100 = 3500 - 17 \times 65 - (93^2 \div 2)$ which would give a negative result as $93^2 \div 2 = 4324.5$ which is already more than 3500. Moreover: If we multiply 35×100 and subtract 17×65 we have $2395 \div 65 = \pm 36.8$ which is obviously less than 0.5. So our answer remains 491.79366.

9.3 Very big roots

At times we read a message of the thirteenth root of a one hundred figure number. I presume that the number one hundred was chosen because it is more impressive than ninety six. In either case the answer will have eight figures.

Basic number	Exp.	Answer	Number of digits
1	20	1	1
2	20	1,048,576	7
3	20	3,486,784.401	10
4	20	1,099,511,627,776	13
5	20	95,367,431,640,625	14
6	20	3,656,158,440,062,976	16
7	20	79,792,266,297,612,001	17
8	20	1,152,921,504,606,846,976	19
9	20	12,157,665,459,056,928,801	20

In the table above you see the numbers 1 up to 9 calculated to the twentieth power and the quantitative difference in the number of digits in the answers. So one cannot simply say the number of digits of the basic number times the exponent gives the number of digits of the answer.

A number of eight figures, calculated to the tenth power can result in an answer of from 71 up to 80 figures: no fewer but also no more.

In the opposite situation: a tenth root of a number from 71 up to 80 figures has in each case an answer of eight figures.

Later on in this chapter you'll find more detailed information about these very big roots, about the reason for the choice of a thirteenth root and no other numbers and why the exponent of this root is always a prime number.

It is not completely clear why the choice was for the 13th root rather than the 11th or 17th. Referring to a saying of Wim Klein "The difficulty is not in the number of the exponent but in the number of figures of the answer". Klein himself extracted a 73rd root of a 500 digit number. For your information, the answer has 7 digits and lies between 6848830 and 7068301.

For the good order: this kind of work is only supported by a very few people, for setting or breaking a world record, and it requires profound training!!

The standard method is finding as many figures as possible by using logarithms; people who commend this operation generally know the logarithms of numbers up to 150 to five or even more decimal places by heart. The remaining figures are found by reasoning.

We have already seen that the thirteenth power numbers each have the same modulo 7 and 13 as the basic number. It is also known that in the thirteenth power the basic number and the result have the same final figure. But there is more.

Exp.1	Exp. 2	Exp. 4	Exp. 5	Exp. 10	Exp. 20
01, 25, 76	24, 49, 51, 74, 75, 99	07, 18, 32, 43, 57, 82, 93	6, 16, 21, 36, 41, 56, 61, 66, 81, 86, 96	04, 09, 11, 14, 19, 29, 31, 34, 39, 44, 59, 64, 69, 79, 84, 89, 91, 94	The rest

Again we use table 1 for more detailed explanation. For the numbers from the columns exp. 1, 2, 5 and 10 we see that in each 10th power it is seen that, e.g. in the 3rd and in the 13th powers, the last two figures of the result are the same.

E.g. $61^3 = 226\,9\,81$; $61^{13} = 16191528743215275565 7\,5\,81$.

If you look carefully you will also see that the hundreds show a significant detail; in the tenth power the tens of the basic number can be multiplied by ten to get the new hundred: 5 81 is 600 more than 9 81. 981 × 601 = (589) 581.

We take another example and compare the final figures of 41^3 and 41^{13}: $41^3 = 68\,9\,21$ and $41^{13} = 9251030231501362 9\,3\,21$. Again we see that the hundreds can be found by multiplying the tens of the basic number by 10: which ends in 321. 921 × 401 = (369) 321.

9.4 Why 13th roots?

Why choose a prime number bigger than ten as the exponent for this type of roots, regardless the magnitude of it? A prime number means that there is only one possibility, which enlightens the reasoning. This is a big advantage, especially with odd numbers. In even numbers there can be four possibilities in the case of three last digits, e.g. 214, 464, 714 or 964.

For one hundred figure question number there is this:

9.4.1 An 11th root

This would mean that the answer will have ten figures, this would be too many. The answer lies between 1,000,000,000 and 1,232,846,739, that's 232,846,740 numbers.

9.4.2 A 13th root

This seems to be a reasonable balance, the lowest possible answer is 41,246,264, the highest possible answer is 49,238,826: so remaining are 7,992,563, almost eight million numbers.

9.4.3 A 15th root

After 13 comes 15, not a prime number but: In the fifth root we have the same final figures as in the first one, and this to the third power leads to a lot of possibilities. E.g. if the last three figures of the question number are 699, from 179 + every 200, all the numbers calculated to the fifteenth power end in 699.

9.4.4 For the 17th root

Here the answer will have "only" six figures, between 666,085 and 762,698, 96,614 numbers

9.4.5 For the 19th root

The answer here also has six figures: between 162,377 and 183,298, so 20,922 numbers.

9.4.6 A 12th root

This would be the combination of a fourth and third root. And with the fourth root if we want to find three figures, there are 16 candidates.

9.4.7 A 16th root

If we want to find four figures, there are 32 candidates. All the enlisted numbers, calculated to the 16th power end on 0001. This looks like looking for a needle in a haystack. If we want three figures there are also 32 candidates.

Now I present you a surprise: the 32 candidates are enlisted and the surprise numbers are in bold: 1, **443, 807**, 1249, 1251, **1693, 2057**, 2499, 2501, **2943, 3307**, 3749, 3751, **4193, 4557**, 4999, 5001, **5443, 5807**, 6249, 6251, **6693, 7057**, 7499, 7501, **7943, 8307**, 8751, **9193, 9557**, 9999.

Indeed; the difference between the bold printed numbers is 364, which is divisible by 91. Therefore extracting sixteenth roots is more like a gamble: it is clear that this kind of question cannot be solved in a reasonable way, as sixteen of the possible answers differ by a multiple of 91.

9.4.8 A 14th or 18th root

The fourteenth and eighteenth roots are the squares of the seventh and ninth roots. These kinds of roots have not yet been used as questions in tournaments. This is very elaborate and requires a profound knowledge of the logarithms plus an extensive knowledge of the changing of the last figures when powering.

The arguments we now have seen make it fully understandable that for very big roots the exponent should be a prime number. If one looks for three figures in the case of an even question number, there are four possibilities and their difference is not a multiple of either 7 and 13, nor is it a multiple of 91.

E.g. 128, 378, 628 and 878; all these numbers calculated to the thirteenth power end in 448. They differ 250 which is not a multiple of any modulus.

9.4.9 A 13th root

The "standard" is a big root of a one hundred figure number, which we work out.

For all clarity, a method of working is described. The question of how anyone can solve this kind of question in even a small number of seconds, is fully unknown to the author. To find an answer, one should find at least the first four figures by means of logarithms. For this we need the first six figures of the question number. To find the other figures of the answer one needs modulo calculation and knowledge of the thirteenth powers.

The solution for this kind of question is found:

- By logarithms for the – from left to right – first four or even five figures
- By reasoning and a profound knowledge of the structure in the powers the last – from right to left – two, three or even four figures
- Modulo calculation
- A very thorough preparation

E.g. the question-number:
5,066,669,315,731,832,474,796,071,519,412,465,480,532,546,732,296,216,464,241,4
13,735,000,308,440,784,776,460,010,919,907,342,271,653

is a one hundred figure question number which is the result of an eight figure number, raised to the thirteenth power.

The logarithm of 506666 = 0.70472. This is the result of interpolation. The log of the complete question number is 99.70472.

We divide the log of the question number by thirteen and then get the logarithm of the answer number, i.e. four/ five decimals $99.70472 \div 13 = 7.66959$, the first digits of the answer are 467 ± 2 or 3.

With the last four figures (1653) of the question number we want to find the last four figures of the answer number. As the last two figures of the question number are 53, the basic number is 13. 13^6 ends with 6809, and the last four figures of 13^{13} are 6809 × 6809 × 13 = 2481 × 13 = 2253. The jump for $13^{13} = 13 \times 13^{12}$ (last two figures) × 100 = 13 × 81 × 100 = 5300 per hundred.

From 2253 – basic number to the power of 13 – up to 1653 we go up by 9400. 9400 ÷ 53 = 98, which means that the last four figures of the answer numbers are 9813.

So far we have found that the answer will be 46 7 – (2/3) – 9813. We have now already seven correct figures of the answer number; what remains is to find the exact fourth one, either a two or a three.

This is the moment to invoke the help of modulo calculation with 91.

For the mod 91 calculation the question number is transferred as follows:

5066 669315 731832 474796 071519 412465 480532 546732 296216 464241 413735 000308 440784 776460 010919 907342 271653

Part	Mod 91 pos.		Mod 91 neg.	
271653	382	18		
907342			565	19
010919	909	90		
776460			316	43
440784	344	71		
000308	308	35		
413735	322	49		
464241			223	41
296216			80	80
546732	186	4		
480532	52	52		
412465	53	53		
071519	448	84		
474796	322	49		
731832	101	10		
669315			354	81
5066	61	61		
		576		264
	Balance	312	= + 39 (mod 91)	

An intermediate step: As 1001 = 0 (mod 91) then 1000 = -1 (mod 91). So however many thousands there are, so we have that many times the -1 (mod 91). So 289,000 – 3 × 91000 = 16000 = -16 (mod 91). Also 1,000,000 = 1000^2 = -1^2 = 1 (mod 91). If we subdivide a number from right to left in groups of six figures, we can determine the mod 91 of that number by adding and subtracting. For example the mod 91 of the number 153 879 349 642 931 = + 21 – 5 + 76 – 60 + 62 = 3 (mod 91).

For eludication: each part of six figures is to be seen as a multitude of 1001 plus or minus something. Example: 271653 = 271 × 1001 + 382 = 18 (mod 91) positive, 907342 = 907 × 1001 – 565 = -19 (mod 91).

The result is that the question number is 39 (mod 91). According to the principle of mod 91 calculation the answer number will be also 39 (mod 91). This is the shape of the answer number: 46 7 – 2/3 – 9 813 of which 46 and 813 are the positive mod 91 parts, 46 + 813 = 859 = + 40 mod 91. Then we get either 729 or 739 as the negative part. We may write these as 729000 and 739000. Which of them is – 1 mod 91? This is 729000.
Now we have the final answer number which is 46 729 813.

Now we do the thirteenth root of a hundred figure number; here the question number is an even number.

This is the question number, find the thirteenth root of this number, magnitude 100 figures.

4055 288514 982105 265158 477536 039719 362968 725868 865642 961603 913116 286634 339556 416730 930375 796413 431808

We take the log of 405528.

Log 405 = 0.60746
Log 406 = 0.60853

The difference between log 405 and 406 = 0.00107. We take this multiplied by 0.00528 and then get 0.0000056496, which we round to 0.00056, add this to 0.60746 and by means of interpolation arrive at 0.60802. As we have a one hundred figure number, the log of it is 99.60802 which we divide by 13 and then get 7.66215. We find 4593 as the first four figures of the answer.

Next, 1808 is from the number 48^{13}, compare with 48^3 which ends with 92.

Part	Mod 91 pos.		Mod 91 neg.	
431808	377	13		
796413			383	19
930375			555	9
416730	314	41		
339556	217	35		
286634	348	75		
913116			797	69
961603			358	85
865642			223	41
725868	143	52		
362968	606	60		
039719	680	43		
477536	59	59		
265158			107	16
982105			877	58
288514	226	44		
4055	51	51		
Total		473		297
		18		24
Balance		-6	+85 (mod 91)	

However, we should not forget that also 98 could be the last digits of our answer!

Now we have to find out the last four figures of 48^{13}, which are 4208. Then the jump of $48^{13} = 13 \times 48^{12} \times 100 = 13 \times 96 \times 100 = 4800$ per hundred. The question number ends with $1808 - 4208 = 7600 \div 48 = 12, 37, 62$ or 87. And therefore the hundreds of our answer will be one of these.

	(mod 91)			(mod 91)
1248	65	+ 1250	2498	41
3748	17	+ 1250	4998	84
6248	60	+ 1250	7498	36
8748	18	+ 1250	9998	79

The shape of the last four figures of the answer number is to be found in the table. Because we have an even number this means that besides the numbers ending in 48 we may also add 1250 to each of them, and then we have to reason which one of them results in the correct answer number.

Our answer so far is 45 930 000 + one of the eight numbers from the table. The check is the modulo 91 of it. 45 represent the millions, so it is + 45 (mod 91). 000 represent units, it is 0. 930 represents the thousands. 930 = 20 (mod 91), in this case -20, the "balance" of 45,930,000 = +25 (mod 91). As the complete answer number will be 85 (mod 91), this means that the four figure number for completing the answer will be 60 (mod 91). In the table you see that 6248 meets this requirement so that the final answer number is 45 936 248. Check: 45 remains 45, 248 = 66 (mod 91); 45 + 66 = 111 minus 936 = 26 (mod 91), 111 − 26 = 85 (mod 91). Q.E.D.

I realise that this chapter is far from simple and I hesitated before writing it. Why do it then? To illustrate that there is nothing magic, it is all strict logic. At times one can hear "Which tricks do you have?". I hope the conclusion of the reader will be that it is not for nothing that the title of this book is "Mental calculation, an art apart". It is a matter of art and not of tricks!

With this description I do not want to suggest that I can solve questions like this, and surely not without the use of paper. It will always be a major performance, which is far beyond my abilities! For me it is not clear how someone can find the solution of this kind of question in a few seconds. Try for yourself how much time it takes just to find the mod 91 of a hundred figure number.

10 Factoring

For the good order: in this chapter we describe factorization as a sport, a challenge, a handicraft, with numbers which are "manageable", generally not bigger than eight figures, which can be factorized without sophisticated computer programs.

One can also say in a more common way that factorization bears a good likeness to a game we all know: the hide and seek game. We know that there are at least two factors in the number given, the question is "where are they"?

Factoring is breaking down a given number in the smallest possible factors, prime numbers. As an example the number 1260 can be factorized as follows: $1260 = 2 \times 2 \times 3 \times 3 \times 5 \times 7$. Formally, a prime number is an integer number having the three properties:

- Greater than 1
- Only divisible by itself and by 1
- An integer number

Apart from the order of the factors, the factorization of a number is unique. The number 1 is not considered as a prime number, because then an infinite number of factorizations is possible, e.g. 15.

$15 = 3 \times 5 = 1 \times 3 \times 5 = 1 \times 1 \times 5 \times 3 \times 1 \times 1 \times 1$ et cetera et cetera and the value of the answer does not change.

One can also say: a prime number has two divisors, 1 and itself, and the number one has only one divisor: itself.

Any even number is divisible by 2, and so 2 is the only even prime number and any even number greater than 2 cannot be prime. A number greater than 1 which is not prime is said to be composite.

In literature we can find heaps of remarkable mental calculation performances, but about factorization and prime numbers there is only a little information.

Concerning myself: on 06–01–2011 as the first one worldwide I set the record for factorisation of five digit numbers, from 10,000 up to 99,999, no prime numbers amongst them. It concerned a set of twenty five digit numbers and the time needed was measured, it was 13 minutes and 38 seconds.

To give an impression, the analysis of the factorisation of five figure numbers:

>From 1–100,000 there are 9,592 prime numbers
From 1–10,000 there are 1,229 prime numbers
Remaining from 10,000 – 100,000 there are 8,363 prime numbers
From 10,000 – 100,000:
Even numbers 45,000
Odd numbers 45,000
Out of them divisible 36,637
Even numbers 45,000
Numbers to work with: 81,637
Even numbers 45,000 = 55.12%
Odd numbers 36,637 = 44.88%

The biggest prime number we see is 49,999 which we get when factoring the number 99,998. From 2 – 49,999 there are 5,134 prime numbers.

The current way of working for factorisation is: calculate the square root of the number offered and divide that number by all the prime numbers smaller than that root. If a division is an integer one, then there is no great labour with the risk of errors. In the remainder, the solution is found. In fact this method is a crude one, which also is described as "trial and error". √100,000 = 316+, by which we only need the 65 prime numbers lower than 316.

Before starting the "tough labour", an anecdote.

On internet I saw a video: "the most stupid girl in the world", unfortunately I did not write down the details. The subject was a quiz, the question was a $50,000 question, the person in question was a girl of ± 20 years of age, with – a perfect example of type casting – long blond hair. She was presented with three titles of films and the question was to point the prime number in one of this titles.

I give you the numbers, the titles are not exactly the same, it concerns only the numbers.

- *The eleven football cracks*
- *The Four Aces Quartet*
- *Police Academy number Six.*

"UUUUUUUUUUUUUUUUUUUUhhhh, that's a nasty question. I know they exist, from school. But I do not remember exactly"
"You may invoke the help of someone in the audience".
"Well, six cannot be a prime number as it is two times three. Eleven is neither divisible by two nor by five, so four is the prime number"

And the poor girl, she followed the advice and pressed the wrong button, and gone was the fifty thousand "box".
But should she be the only one worldwide?

10.1 Smaller prime numbers

On the grounds of the definition of the prime numbers they are in fact the building stones of the number world, as any other number than a prime number is said to be composite or composed. Nevertheless, if you would stand with a blackboard on a square somewhere in a town, and ask the passers-by to write down some prime numbers, there is a great chance that you would not need an eraser.

However, long, long ago there lived a man who had found a very instructive, interesting and challenging method for finding the prime numbers up to 100.
His name was Erastosthenes, born in 276 B.C. and was later the Director of the Library in Alexandria.

The method he found is still known as the "Sieve of Erastosthenes" and works as follows: on paper or a blackboard are written all the numbers up to 100, and by elimination of all the divisible numbers, the prime numbers remain.

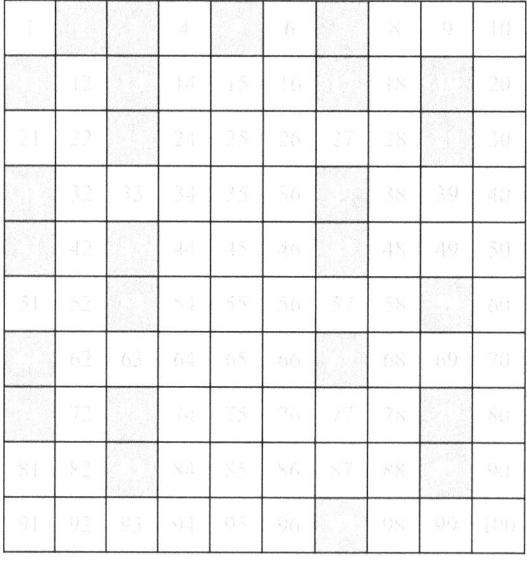

This is the way of working:

Give a stripe through all the numbers which are divisible by 2
Give a stripe through all the numbers which are divisible by 3
Give a stripe through all the numbers which are divisible by 5

Give a stripe through all the numbers which are divisible by 7

And what remains are all the twenty five prime numbers up to 100, there are 25: 2, 3, 5, 7, 11, 13, 17, 19, 23, 29, 31, 37, 41, 43, 47, 53, 59, 61, 67, 71, 73, 79, 83, 89, and 97.

At school one type of question never fails; factorisation, in which the students get presented some numbers, mostly not bigger than 1,000, which have to be factorised. The prime numbers generally do not exceed 20.

A remark

On a website for brain activities appeared the question "How can I learn to memorise all the prime numbers up to 10,000". I am no memory acrobat, and see and read about their remarkable performances. I learn only things by heart if I have a special goal for it and use the knowledge intensively. Therefore my answer: "If this knowledge is not used intensively then what will happen is what always happens in comparable situations: "If you don't use it, you lose it".

10.2 Some simple tricks (and not more than that)

Here follow some tips and simple tricks, although I hate the word profoundly. They are indeed simple tricks and only adaptable for smaller numbers! From students of the pedagogical academy I heard that what I told about the numbers 13, 17 and 19 was completely new for them.

2. All the even numbers contain at least one time 2, so you continue until you get an odd number

3. If the sum of the individual figures in a number is divisible by 3, then the whole number is: $174 = 1 + 7 + 4 = 12$ which is divisible by three, so the whole number is divisible by 3.

5. If a number ends of 0 or 5, the number is divisible by 5.

7. To find out if a number is divisible by a given other number, one can subtract any multiple of the divisor and then divide. Do you want to know if 273 is divisible by 7? Then subtract 210, a multiple of 7, and get 63, which is divisible by 7, so 273 is too. And that the result is 39, which is very simple. Or: split 273 in 2 and 73. Then $2 \times 2 = 4 + 73 = 77$ which is divisible by 7. What we do here is use the fact that $100 = 14 \times 7 + 2$. $1673 = 73 + 2 \times 16 = 105 = 5 + 2 \times 1 = 7$, so 1673 is divisible by 7.

11. Subdivide the given number from right to left in groups of two figures and add the groups. If the result of this addition is divisible by 11, then the whole number is.

2167→ 21 + 67 = 88, which is divisible by 11, so 2167 is too. When this is known, then subtract a multiple of 11, e.g. 1100, then what remains is 1067. Then subtract 77, and you get 990, the final result of 2167 ÷ 11 = 197.

13. Consider that 8 × 13 = 104, 100 + 4. Use this as follows: We take 156 which we subdivide → 1 and 56. Now take 1 and multiply by 4 and get 4. Subtract 56 − 4 = 52, which is divisible by 13, so 156 is too. And 173? 73 − 4 × 1 = 69, which is not divisible by 13. And if we divide 173 ÷ 13 we now know that the remainder is 4.

17. The same idea as with 13, with this difference: 6 × 17 = 102, so 100 + 2. We subdivide the number 238 into 2 and 38. Now we do 2 × 2 = 4 and subtract 38 − 4 = 34, which is divisible by 17, so 238 is as well. And 1819? Well, 19 − 2 × 18 = -17 which is divisible by 17, so 1819 is too. And 2347? 47 − 2 × 23 = 1, so 2347 is not divisible by 17. The remainder of this division is 1.

19. Again the same idea. 5 × 19 = 95, or 100 − 5. This means that the given number is treated as follows: 266 becomes 2 66. Now we do 5 × 2 = 10 and ADD, because 95 is less than 100, 66 + 10 = 76, which is divisible by 19, so 266 is too. And 1942? 5 × 19 = 95 + 42 = 137, which is not divisible by 19. The remainder of the division is 4.

I met a boy of eleven years who was surprisingly good in mental calculation, he gave only correct answers. I had a short talk with his mother, who knows that I am good in prime numbers. "Do you happen to know some prime numbers?" "Yes, 2, 3, 5, 7, 11, 13, 17, 19" without any hesitation. After 20 it was a bit slower, but she came to 37. I fear, a high exception.

10.3 The greater work

In the book " The Great Mental Calculators" by Steven B. Smith we find 3 names of famous mental calculators who were cracks in factorization:

- Zerah Colburn, 1804 – 1840
- Zacharias Dase, 1824 – 1861
- Wim Klein, 1913 – 1986

About Wim Klein we know that he liked factorization very much, unfortunately exact performances are not mentioned. Nor do we know of any documents in which he describes his methods.

Present names: Willem Bouman – 1939; Jerry Newport – 1948; Wenzel Grüß – 2002.

In May 2012 Bouman did another remarkable performance: "four consecutive primes". That is numbers from 13 – 16 figures are the product of four consecutive 4 digit prime numbers. E.g. 4565480435215321 = 8209 × 8219 × 8221 × 8231. Details about this are to be found also on the website recordholders.org.

10.4 Methods

The current working methods for factoring are:

By taking the square root of the given number and dividing the number by all the prime numbers smaller than this root. Since, $\sqrt{100{,}000} = 316+$, we need only the 65 prime numbers up to 316, of which in this case is the number 313. If a division is exact, i.e. there is no remainder, we have found a factor. This means a huge quantity of elementary computations, with the risk of errors. In fact this a rough technique, which generally is called the "trial and error" method.

Examining if the given number can be written twice as the sum of two different squares. If so there are two prime factors. Further on in this book this will be worked out.

Examining if the given number can be increased to an exact square, by adding another square, in the algebra well known as the remarkable product: $a^2 - b^2 = (a + b) \times (a - b)$. More about this later on.

10.5 Tools

As tools for factoring I use:

- My profound knowledge of all the multiplications of the two digit numbers
- My insight into numbers
- The filter methods which I elaborate in this book
- My ready knowledge of the prime numbers up to 10,000

My attention was focused to develop filter methods which reduce the mechanical labour, to work faster and to be more accurate. It is also to be realiszed that the bigger the numbers get, the more time the factorization will take.

10.6 Filtering methods

- There are the numbers which can quickly be filtered out by mechanical labour: 2, 3, 5, 7, 11, 13. Anyone who commands the 2 × 2 multiplications can start with this and the primes up to 100 will bring no skilled mental calculator any trouble.
- Being about 14 years old I discovered the structure in the squares. Between

1–100 the squares of numbers that do not end in 0 or 5 can be subdivided in groups of four numbers with two identical final digits.

Knowledge of the squares is a very useful expedient, not only for finding them quickly when factoring, moreover an insight into the sums and differences of squares – more about them later – can accelerate the labour of factorization considerably.

10.7 Something about the squares

The American author of numerous books about mental calculation and mathematics Ron Doerfler wrote this about me in one of his articles:

Your knowledge of squares must be very useful for multiplying large numbers as well. In things I've written, I've pointed out that knowing squares is the single best thing for mental calculation generally, and you must be the world expert in that!

Of course it is very pleasant to read these words, but there is more: again it appears that there is an intensive interlink between all the arithmetic operations.

As a boy of about 14 years I discovered the structure in the squares.

Table 1

2^2	4	3^2	9	7^2	49	8^2	64
48^2	2304	47^2	2209	43^2	1849	42^2	1764
52^2	2704	53^2	2809	57^2	3249	58^2	3364
98^2	9604	97^2	9409	93^2	8649	92^2	8464

Between 1 and 100 the squares can be subdivided in groups of four numbers with the same final figures. It is easily to be seen that the sums of the first and last one, also number 2 and 3, is always 100.

Table 2

11^2	121	23^2	529
239^2	57121	227^2	51529
261^2	68121	273^2	74529
489^2	239121	477^2	227529
511^2	261121	523^2	273529
739^2	546121	727^2	528529
761^2	579121	773^2	597529
989^2	978121	977^2	954529

In table 2 you see the same symmetry, now with numbers of three figures.

Playing with numbers and squares as a boy of 15 I made a remarkable discovery: If a number can be written as twice the sum of two different squares, this number can be factorized in two factors, each of which in itself is also the sum of two different squares. Also I found that if a number can in only one way be written as the sum of two different squares, it is a prime number. In my thinking there are two kinds of prime numbers: the usual ones like 3, 7, 11, 19, et cetera all 3 (mod 4) and the special ones which are the sum of two different squares: 5, 13, 17, 29, 37 et cetera, all 1 (mod 4).

It is known that if a number is in one or more ways the sum of two different squares, it is by definition 1 mod 4. Also no number 3 (mod 4) can be written as the sum of two different squares, 1 even and 1 odd. Formulated more sharply: If a number is 1 (mod 4) and it is one time the sum of two different squares and the number is not divisible by the square of a number which is 3 (mod 4), then this number is prime.

Because this book also will be read by mathematicians, I have had it read by Dr. B.M.M. de Weger, a mathematician and assistant professor at the Technical University of Eindhoven. He brought me the disillusioning news that the mathematician Pierre de Fermat in 1640 already considered that every number 1 (mod 4) is the sum of two different squares, but he did not produce the proof of it. (We find this also in the presentation of Prof. Aitken in November 1954 at the university of Edinburgh). The first proof published comes from Leonard Euler in 1749. However, it is the question if these mathematicians hereby also have thought on factorization, there is nothing to be found about this in published literature.

Between 10,000 and 100,000 there are 5,052 odd numbers which can be written as the sum of two different squares in at least two different ways. Out of them 2,215 are less than 50,000. E.g. 10,001 = $100^2 + 1^2$ and $76^2 + 65^2$. As soon as a composition like this has been found, the factors are quite simple to derive: $(100 \pm 76) \div 2 = 88$ and 12. $(65 \pm 1) \div 2 = 33$ and 32. 88 and 33 have 11 as a common factor, resp. 8× and 3×. 32 and 12 have 4 as a common factor, resp. 8× and 3×. Then the factors of 10,001 are $11^2 + 4^2 = 137$ and $8^2 + 3^2 = 73$. Nota bene: it is a condition that the BIGGEST common divisor is taken!

To be able to execute these operations a profound knowledge of the squares is a necessity.

Over 55 years I lived in the illusion that the above mentioned was strictly new and I looked for a favorable occasion to publish it. And in silence I was very proud of my "invention".

In relation to the above mentioned I discovered that if one takes the number of special primes being N, and multiplies them, for the product it is then so that then the number of possibilities of sums of squares for combination is $2^{(n-1)}$. Four of this kind of special prime numbers give 2^3 possibilities for combination e.g.: $5 \times 13 \times 17 \times 29 = 32045$ which can be composed with the sum of the following pairs of squares: {179 + 2}, {178 + 19}, {173 + 46}, {166 + 67}, {163 + 74}, {157 + 86}, {142 + 109}, {131 + 122}. This is only valid when all prime factors are special primes. If one of the special primes is squared there are not four, but three combinations possible and one combination is a multiple of the number which is squared. E.g. $325 = 1^2 + 18^2 = 17^2 + 6^2 = 15^2 + 10^2$.

This was also already known by Diophantes in the 3rd century.

It is also of paramount importance to know which combinations are possible and which ones are not. Consider that up to 100 there are 21 rows of squares with different final digits. The numbers ending on 5 have squares ending on 25, and those ending in 0 have squares ending in 00. By definition squares are either 0 (mod 4) or 1 (mod 4). Therefore the sums of two squares can only be 0, 1 or 2 (mod 4) and never 3 (mod 4). The table below is a good tool to see which combinations are possible or impossible.

A useful expedient to see how the sum of squares is composed is modulo 9 calculation. If a number offered is 0 (mod 9) there is no other possibility than each of the squares is 0 (mod 9). The basic numbers 0 up to 9 squared are resp. 0, 1, 4, 0, 7, 7, 0, 4, 1 and 0 (mod 9). If the sum of 2 squares is 1, 4 or 7 (mod 9) then automatically one of these squares is 0 (mod 9). Numbers 2, 5 of 8 (mod 9) are composed of sums of squares 1, 4 or 7 (mod 9) in any combination.

This technique I consider as the first filter. By a 'filter' I mean a technique to reduce the theoretical number of combinations possible to find the right combination.

As in table 3 we work with hundreds, and modulo 16 will be used. Table 3 below considers all possible values for the tens and units, and states whether the corresponding value in the hundreds column is odd or even.

Table 3

Final digit	Tens	(H)undreds (E)ven / (O)dd
1	01	Always even
	21	Always odd
	41	Always even
	61	Always odd
	81	Always even
4	04	E with H 0 (mod 4) and O with H 3 (mod 4)
	24	E with H 2 (mod 4) and O with H 3 (mod 4)
	44	E with H 6 and 14(mod 16) and O with H 1 and 13 (mod 16)
	64	E with H 0 and 4 (mod 16) and O with H 1 and 9 (mod 16)
	84	E with H 4 and 12(mod 16) and O with H 3 and 7 (mod 16)
5	25	Always even, H never 4 of 8
6	16	E with H 0 and 12(mod 16), O with H 5 and 13 (mod 16)
	36	E with H 0 and 8 (mod 16) and O with H 3 and 15 (mod 16)
	56	E with H 2 and 6 (mod 16) and O with H 3 and 11 (mod 16)
	76	E with H 6 and 14 (mod 16) and O with H 5 and 9 (mod 16)
	96	E with H 8 and 12 (mod 16) and O with H 1 and 9 (mod 16)
9	09	Always even, H all E (mod 16): 0, 2, 4, 6, 8, 10, 12, 14
	29	Always odd, H all O (mod 16): 1, 3, 5, 7, 9, 11, 13, 15
	49	Always even H all E (mod 16): 0, 2, 4, 6, 8, 10, 12, 14
	69	Always odd, H all O (mod 16): 1, 3, 5, 7, 9, 11, 13, 15
	89	Always even H, all E (mod 16): 0, 2, 4, 6, 8, 10, 12, 14

10.8 Examples of sums of squares

8641: 1 (mod 9). One of the squares has to be 0 (mod 9) and the other one 1 (mod 9). 41 can be composed by adding xx16(EH) + 25 (always even hundreds) or xx41 (even hundreds) + 00 (even hundreds). $4^2 = 7$ (mod 9) so it drops out, $96^2 = 9,216 => 8,641$, this drops out too. $21^2 = 441$, $8,641 - 441 = 8,200$ which is no square, $29^2 = 841 = 4$ (mod 9) and so drops out too. $71^2 = 5,041 + 3,600 = 8,641$. A full hit! There are no other possibilities for combination, so 8,641 is prime. For the good order: 79^2 ends indeed in 41 but drops out as it is 4 (mod 9).

6953: 5 (mod 9). (OH)53 is to be composed as follows: 09 (always EH) + 44 (OH), 29 (always OH) + 24 (EH), 64 (EH) + 89 (always EH) or 84 (OH) + 69 (always OH).

The number of possibilities for choice can be reduced considerably by doubling the number. 13,906 = 1 (mod 9) and can only be composed by xx81 (always EH) and xx 25 (always EH) where one of the squares is 0 (mod 9) and the other one is 1 (mod 9). $9^2 = 81$, but 13,825 cannot be a square. 41^2 and 59^2 are 7 (mod 9) so drop out. $91^2 = 8,281 = 1$ (mod 9).

13,906 − 8,281 = 5,625 = 75^2, so it is a hit. Furthermore we find $109^2 + 45^2$. We now do retrospective work, by dividing by 2. Then (91 ± 75) ÷ 2 equals 83 and 8, and (109 ± 45) ÷ 2 equals 77 and 32. Then we deal with the odd numbers; (83 ± 77) ÷ 2 and then get 80 and 3. For the even numbers, (32 ± 8) ÷ 2 and then we get 20 and 12. The biggest common factor of 80 and 20 = 20, resp. 4× and 1×. The biggest common factor of 12 and 3 is 3, resp. 4× and 1×. So the factors of 6,953 are $20^2 + 3^2 = 409$ and $4^2 + 1^2 = 17$.

We now have had two filters: modulo 9 and Table 3.

To reduce the quantity of work to be done it is recommended to divide the given number firstly by all the primes < 100. This can be done very quickly because of the small numbers and the rate of errors is minimal. In theory there are 25 divisions out of which the primes 2 to 19 almost automatically drop out.

We start to work with the first prime number >100: 101. In theory this is the smallest factor, the biggest one is 991 as 101 × 991 = 100,091. Furthermore 101 + 991 = 1,092. This is the maximum sum of two prime factors. The maximum square we have to deal with will be the half of it: 546.

The biggest square we have to add to the number offered will not need to add to more than $546^2 = 298,116$. In theory this is $445^2 = 198,025$, because 100,091 is 2 (mod 9) which number only can be completed to a square with a square 7 (mod 9). The biggest possible prime factor is 313, after all $317^2 = 100,489$. Between 101 and 313 there are 40 primes.

39493. The number is 1 (mod 9). Concerning the sum of squares only one combination is possible: 0 + 1. But 93 can be obtained in many ways: 04 + 89; 09 + 84; 24 + 69; 29 + 64; 44 + 49. If we multiply the number by 2 we get 78,986, of which the attractive aspect is that here is only one solution possible: 61 (always OH) + 25 (always EH). But 78986 = 2 (mod 9), so in the light of mod 9 the following combinations are possible 1 + 1 and 4 + 7. Firstly we exclude all the squares 0 (mod 9), so all those of which the basic number is divisible by 3. Squares ending with 161 drop out as no square ends on 825. Squares ending on 561 also drop out as there is neither a square ending on 425.

We can start with 19 and 31, (69, 81, 119, 131 and 169 drop out as described here above). Next we have 181, 269 and 281. 78986 − (181^2 =) 32761 gives 46225, the square of 215. Hurray, we have already one combination. 78986 − (269² = 72361) = 6625, which immediately drops out: 6625 does not exist as a square. 78986 − (281² = 78961) = 25 = 5².

Okay, we now have two combinations: 181² + 215² and this one: 281² + 5². Because we doubled 39493 to 78986, we now undo this and derive the squares: (215 ± 181) ÷ 2 gets 198² + 17² and (281 ± 5) ÷ 2 gets 143² + 138². Next, the even numbers give us (198 ± 138) ÷ 2 to get 168 and 30. Next (143 ± 17) ÷ 2 get 80 and 63.
The biggest common factor of 168 and 63 is 21, respectively 8 × and 3 ×. The biggest common factor of 80 and 30 is 10, respectively 8 × and 3 ×.

So the factors of 39,493 are 8² + 3² = 73 and 21² + 10² = 541.

Example 88183. As this number is 3 (mod 4) it cannot be the sum of two different squares. √88,183 = 296.9. The biggest prime number < 297 = 293. The primes up to 100 are used for division. From 101 up to 293 there remain 36 prime numbers. One should be aware that the risk of errors is much greater when dividing by larger numbers.

Our first conclusion is that one of the factors of 88.183 must lie between 101 and 293. This is selection one.

Because we know that 88,183 is not prime, given the problem, we can use the formula (a² − b²) = (a + b) × (a − b) where 88,183 represents (a + b) × (a − b), to which number we have to add b² to obtain a².

The smallest factor, a − b, is > 101 because we have tried all factors up to 100. If 101 should be the smallest possible factor, 877 will be the biggest one: 101 × 877 = 88,577. This is selection two.

(101 + 877) ÷ 2 = 489² = 239,121 − 88,183 = ± 151,000. 388² = 150,544, the square we are looking for then lies between 1 and 388. This third selection brings us this: 88,183 = 1 (mod 9) so that the square we have to add herewith has to be 0 (mod 9) and by consequence the basic number of it will be 0 (mod 3).

From 0 up to 9 the modulo 9 squares of the successive numbers are: 0, 1, 4, 0, 7, 7, 0, 4, 1, 0.

The fourth selection: The square required has to end in 1 and even tens, after all the other final digits of the squares – 0, 4, 5, 6, 9 – added to the 3 of the number offered would lead to a number which cannot be a square. A square ending in 6 has automatically an odd ten – 16, 36, 56, 76 or 96. Adding such a square to 88,183 would give a number ending in 9 with an odd ten. Such a number can never be a square, see table 3.

So the following numbers match the criteria: 9, 21, 39, 51, 69 up to 381, so with jumps of 30. So up to 381 there are 26 candidates.

Now we have found that none of the factors is < 101, the difference between the number offered and $101^2 = 10201$ is ± 78,000 which means that the square to be found will not lie more than 78,000 above 88,183; so that 279 is the highest number we need to try. This is selection 5.

As the sixth selection we now invoke the help of modulo 16, as 16 has no common divisors with 9 and modulo 4 offers too few possibilities for composing squares. E.g. 88,183 = 7 (mod 16), if we add to that a square which is 9 (mod 16) we have a square with basic number 0 (mod 4), with which we can reduce the number of possibilities rather than a number which is 2 (mod 4). The squares of the numbers from 1 to inclusive 15 are respectively 1, 4, 9, 0, 9, 4, 1, 0, 1, 4, 9, 0, 9, 4, 1 (Mod 16).

Table 4 gives us the mod 9 and mod 16 properties of the squares up to 72. Each number mentioned in table 4 can be increased by 72 until the completion wanted is obtained.

Table 4

Moduli of the squares: Mod 9: 0, 1, 4, 0, 7, 7, 0, 4, 1. Mod 16: 0, 1, 4, 9, 0, 9, 4, 1, 0, 1, 4, 9, 0, 9, 4, 1.							
Rule nr	Mod. 9 resp. 16	Basic number of squares possible					
1	0–0	12	24	36	48	60	
2	1–0	8	28	44	64		
3	4–0	16	20	52	56		
4	7–0	4	32	40	68		
5	1–1	1	17	55	71		
6	1–4	10	26	46	62		
7	1–9	19	35	37	53		
8	4–0	16	20	52	56		
9	4–1	7	25	47	65		
10	4–4	2	34	38	70		
11	4–9	11	29	43	61		
12	7–0	4	32	40	68		
13	7–1	23	31	41	49		
14	7–4	14	22	50	58		
15	7–9	5	13	59	67		
16	0–1	9	15	33	39	57	63
17	0–4	6	18	30	42	54	66
18	0–9	3	21	27	45	51	69

88183: 1 (mod 9) and 7 (mod 16) which includes that b^2 has to be 0 (mod 9) and 9 (mod 16) according to table 10.

To provide a good insight into the squares, we now have table 5.

Table 5

Tens	Units	To complete to				
		With	To	With	To
Even	1	00	01	04, 24, 44, 64, 84	25
	3	01, 21, 41, 61, 81	04, 24, 44, 64, 84		
	7	09, 29, 49, 69, 89	16, 36, 56, 76, 96		
	9	00	09, 29, 49, 69, 89	16, 36, 56, 76, 96	25
Odd	1	25	16, 36, 56, 76, 96	09, 29, 49, 69, 89	00
	3	16, 36, 56, 76, 96	09, 29, 49, 69, 89		
	7	04, 24, 44, 64, 84	01, 21, 41, 61, 81		
	9	25	04, 24, 44, 64, 84	01, 21, 41, 61, 81	00

The seventh selection is the squares 1 (mod 9) and 0 (mod 16), the basic numbers therefore being 1 or 8 (mod 9) and 0 (mod 4), < 88,183, and not bigger than ± 166,000. The possibilities are 332, 352, 368, 388. We start with $332^2 = 110,224 - 88,183 = 22,041$. Here the numbers 51, 69, 99 and 141 drop off, as their squares are each < 22,000.

So we can start with $+ 171^2 = 29,241 + 88,183 = 117,424$, which is no square. $88,183 + 189^2 = 35,721 + 88,183 = 123,904$, this is a "full" square, 352^2. We now have found the solution. $88,183 = (352 + 189) \times (352 - 189) = 541 \times 163$.

Another example: A number ending on xx17 can be completed with a number yy04 to a number zz21: $23,617 + 48^2 = 25,921 = 161^2$. Or xx17 + yy44 = zz61: $44,117 + 3,844(62^2) = 47,961 = 219^2$. Table 6 gives some more concrete applications.

Table 6

Examples	+	Gives	= ²	+	Gives	
2,101	8,100	10,201	101^2			
6,901	324	7,225	85^2			
943	81	1,024	32^2			
4,307	49	4,356	66^2			
9,709	900	10,609	103^2			
21,409	9,216	30,625	175^2			
3,111	25	3,136	56^2	6,889	10,000	100^2
3,713	256	3,969	63^3			
7,117	101,124	108,241	329^2			
22,019	4,225	26,244	162^2			
60,279	1,225	61,504	248^2	151,321	211,600	460^2

For enlightenment, another example:

24779: 2 (mod 9) and 11 (mod 16). Also this number is 3 (mod 4), so not the sum of two different squares. The square to be added must be 7 (mod 9) and 9 (mod 16) and can be xx21, together yy00 (OH) or xx25, together yy04 (EH)). We then have a square 4 (mod 16) so the basic number will be an even number, not divisible by 4. Xx21 candidates have to be with the hundreds 1, 3 or 7 as the hundreds, 5 and 9 lead to hundreds which never can be a square: 300 or 700. Xx21 Candidates 139, 239, 311. Xx25 Candidates 05, 85, 275, 355, 365. $\sqrt{24{,}779} = 157+$, so the smallest factor lies between 7 and 157. By quick divisions we can cast out the numbers up to 50. If 53 were the smallest factor, its counterpart should be $\pm$ 460, which in principle is possible. We can now select the candidates which meet the requirements; 7 (mod 9) and 9 (mod 16) and lie between 53 and 157.

Concerning xx25 we have 05, 85, 275, 355, 365, which squared and added to 24,779 give: 24,804, 32,004, 158,004, and 222,804, none of them being squares. Concerning xx21 gives 139^2, immediately a full hit as $24{,}779 + 19{,}321 = 44{,}100 = 210^2$. So the factors of 24,779 are 210 ± 139; 71 and 349.

Very shortly I discovered another technique, on which I am very proud.

Thinking about and playing with numbers, to my big joy and surprise I discovered the following. If a number cannot be written as twice the sum of two different squares at all and it is 1 (mod 4) then this number can be factorized into two numbers, each of them being 3 (mod 4). My theoretical knowledge is insufficient to give an explanation for this. In this relation it is pleasant to remark that Robert Fountain, a promoted nuclear physicist, described my activities with the nice term "number practitioner". Because I do not like to cheat myself, I tested my "invention" a big number of times. Because in every case the correctness of my operation was proved, I keep it for certain.

Here again we see a nice difference between the world of thought of the mathematicians and my one. Immediately after I have made a certain statement, the reaction is "But Mr. Bouman, things like that you have at least to prove". And of course I cannot, as I have never studied number theory. But I found a good defence; "Please, you should prove that it is incorrect what I am doing". And until now such a proof is not presented to me.

But also here a disillusioning surprise awaited me: Dr. De Weger let me know that also this matter had been discovered in the past by the mathematicians De Fermat and Euler. In the literature however there is no evidence that these mathematicians had thought on a technique for factorization.

10.9 Some more examples

$301 = 7 \times 43$, $437 = 19 \times 23$, $22{,}493 = 83 \times 271$, $95{,}477 = 307 \times 311$.

Finally I answer a (possibly already raised) question: " Why must the factorization world record concern five digit numbers?". Although they are not acknowledged as these by the mental calculators, there exist many "memory acrobats". These are persons with an enormous capacity of memory; there is thought on more than 400,000 numbers, amongst which some have up to 200 digits per number. Out of this enormous capacity, answers are reproduced and if necessary a bit of calculation. The "average" audience cannot discriminate the difference, lightning speed is seen and a correct answer is presented – and this is enough. As there is the possibility that such a memory expert learns by heart the factorization of all four digit numbers, the value of a record would seriously be reduced. Learning by heart the factorization of all the five digit numbers is deemed to be impossible.

In 2006, owing to my knowledge and experience with the prime numbers, I was the only mental calculator who scored in a tournament the full 100 points in factorisation, the others scored only 25, 12.5 and even 0.00 – Robert Fountain, mentioned before, gave me the very nice epithet "William Flash, King of the Primes". As far as I know at this time the American Jerry Newport has also a great knowledge of and experience with the prime numbers and factorization. And we should not forget the young German boy Wenzel Grüß (2002), who broke the world record factorization in September 2016.

The last 3 pages of this article contain tables beginning with 0 mod 9 up to 8 mod 9 in combination with mod 16. They will help you to find a completing square for the number you are working with. E.g. you have a number 2 (mod 9) and 7 (mod 16), e.g. 119. In table 11 you'll find suggestions for the square to complete to another square you are looking for. For 119 you can take 5^2 to complete to 12^2. 407 has the same properties: It is 2 (mod 9) and 7 (mod 16). So the "little" square to complete has to be 7 (mod 9) and 9 (mod 16), and has to end with 9. Therefore the basic number will be either 4 or 5 (mod 9) and has to end with 3 or 7. As a consequence, the basic number of the "big" square will be divisible by 3 and 4, so it is a multiple of 12, ending on either 4 or 6. For the "little" square we can think on this: 13, 23, 67, 77. For the big square there are 24, 36, 84, 96. Adding 13^2 gives $407 + 169 = 576 = 24^2$, the final answer about the factors of 407 is $24 - 13 = 11$ and $24 + 13 = 37$.

In the Netherlands, the handy numbers have 8 figures. Amongst them undoubtedly there will be very "nasty" numbers; if there is some recognition, they can be factorized. For this there are neither special rules nor special techniques.

E.g. 29528184. To be split into 2952 and 8184. Each 4 digit number is divisible by

24, but not both by 44; only 8184 is. 29528184 ÷ 24 = 1230341. Splitting results in 1230 and 341. These numbers differ 889, so 1,230,341 is divisible by 7. The result here is 175,763. Both 175 and 763 are divisible by 7, the answer of 175763 ÷ 7 = 25109. As 25 and 109 differ 84, the complete number is divisible by 7, answer 3587. This number is not divisible by 2, 3 or 5, moreover even not by 7, as 35 is divisible by 7 and 87 is not. As 35 + 87 = 122, which is not divisible by 11, this prime number drops out. Then we look at 13: as 300 × 13 = 3900, − 13 = 3887, the next lower number ending on 87 divisible by 13 will be 2587; we thus conclude that 3587 is likewise not divisible by 13. We now do 87 − 2 × 35 = 17, so 3587 is divisible by 17, 211 times. 211 is a prime number, so the complete factorization of 29528184 = 2^3 × 3 × 7^3 × 17 × 211.

12831141. To be split in 12831 and 141, as 91 × 141 = 12831 and 141 = 3 × 47. Now we see that 12831141 ÷ 141 = 91001.

- This number can be written as $301^2 + 20^2$ and $299^2 + 40^2$.
- (301 ± 299) ÷ 2 give 300 and 1
- (40 ± 20) ÷ 2 give 30 and 10
- The common factors of 300 and 30, on the other hand are 10 and 1.
- One of the factors then is $10^2 + 1^2 = 101$.
- Another factor is $30^2 + 1^2 = 901$
- 901 = $30^2 + 1^2$ and $26^2 + 15^2$
- (30 ± 26) ÷ 2 = 28 and 2.
- (15 ± 1) ÷ 2 = 8 and 7. Common factors of 28 and 7 = 7, times 4× and 1×
- Common factors 8 and 2 = 4, times 4× and 1×
- The factors of 901 are $7^2 + 2^2$ = 53 and $4^2 + 1^2$ = 17

Finally 12831141 = 3 × 17 × 47 × 53 × 101.

Table 9

Mod 9		Mod 16		Ex.	+ (mod 9)		(mod 16)		Possible	Example
					To complete with					
0	(mod 9)	0	(mod 16)	144	1,4,7	(mod 9)	1,4,9	(mod 16)		$144+35^2=37^2$
0	(mod 9)	1	(mod 16)	81	1,4,7	(mod 9)	0	(mod 16)	12,4	$81+12^2=15^2$
0	(mod 9)	2	(mod 16)	18		(mod 9)		(mod 16)		
0	(mod 9)	3	(mod 16)	99	1,4,7	(mod 9)	1	(mod 16)	1,15,49	$99+49^2=50^2$
0	(mod 9)	4	(mod 16)	180	1,4,7	(mod 9)	0	(mod 16)	8,44	$180+44^2=46^2$
0	(mod 9)	5	(mod 16)	117	0	(mod 9)	4	(mod 16)	2,18	$117+2^2=11^2$
0	(mod 9)	6	(mod 16)	54		(mod 9)		(mod 16)		
0	(mod 9)	7	(mod 16)	135	0	(mod 9)	9	(mod 16)	all 0 (mod 3)	$135+3^2=12^2$
0	(mod 9)	8	(mod 16)	72	0	(mod 9)		(mod 16)	9,15,33,39,57,63	$72+3^2=9^2$
0	(mod 9)	9	(mod 16)	153	0	(mod 9)	1,4,9	(mod 16)	4,24,76	$153+4^2=13^2$
0	(mod 9)	10	(mod 16)	90		(mod 9)		(mod 16)		
0	(mod 9)	11	(mod 16)	27	1,4,7	(mod 9)	1,4,9	(mod 16)	3,13	$27+13^2=14^2$
0	(mod 9)	12	(mod 16)	108	1,4,7	(mod 9)	4	(mod 16)		$28+6^2=8^2$
0	(mod 9)	13	(mod 16)	45	1,4,7	(mod 9)	4	(mod 16)	6,22	$45+6^2=9^2$
0	(mod 9)	14	(mod 16)	126		(mod 9)		(mod 16)		
0	(mod 9)	15	(mod 16)	63	1,4,7	(mod 9)	1	(mod 16)	9,31	$415+39^2=44^2$

Table 10

Mod 9		Mod 16		Ex.	+ (mod 9)		(mod 16)		Possible	Example
					To complete with					
1	(mod 9)	0	(mod 16)	64	0	(mod 9)	all	(mod 16)	all 0 (mod 3)	$64+6^2=10^2$
1	(mod 9)	1	(mod 16)	1	0	(mod 9)	0	(mod 16)	6,12,18,24,36,48	$145+12^2=17^2$
1	(mod 9)	2	(mod 16)	82	0	(mod 9)		(mod 16)		
1	(mod 9)	3	(mod 16)	19	0	(mod 9)	1	(mod 16)	9,15,33,39,57,63	$451+15^2=26^2$
1	(mod 9)	4	(mod 16)	100	0	(mod 9)	0	(mod 16)	12,24,36,48,60	$100+24^2=26^2$
1	(mod 9)	5	(mod 16)	37	0	(mod 9)	4	(mod 16)	6,18,30,42,54,66	$325+6^2=19^2$
1	(mod 9)	6	(mod 16)	118	0	(mod 9)		(mod 16)		
1	(mod 9)	7	(mod 16)	55	0	(mod 9)	9	(mod 16)	all 0 (mod 3)	$775+3^2=28^2$
1	(mod 9)	8	(mod 16)	136	0	(mod 9)		(mod 16)	9,15,33,39,57,63	$136+33^2=35^2$
1	(mod 9)	9	(mod 16)	73	0	(mod 9)	1,4,9	(mod 16)	12,24,36,48,60	$217+12^2=19^2$
1	(mod 9)	10	(mod 16)	10	0	(mod 9)		(mod 16)		
1	(mod 9)	11	(mod 16)	91	0	(mod 9)	1,4,9	(mod 16)	0	$91+3^2=10^2$
1	(mod 9)	12	(mod 16)	28	0	(mod 9)	4	(mod 16)	6,12,18,24,36,48	$28+6^2=8^2$
1	(mod 9)	13	(mod 16)	109	0	(mod 9)	4	(mod 16)	6,12,18,24,36,48	$253+6^2=17^2$
1	(mod 9)	14	(mod 16)	46	0	(mod 9)		(mod 16)		
1	(mod 9)	15	(mod 16)	127	0	(mod 9)	1	(mod 16)	9,15,33,39,57,63	$415+39^2=44^2$

Table 11

Mod 9		Mod 16		Ex.	+ (mod 9)		(mod 16)		Possible	Example
					To complete with					
2	(mod 9)	0	(mod 16)	128	7	(mod 9)	0	(mod 16)	4,32,40,68	$128+14^2=18^2$
2	(mod 9)	1	(mod 16)	65	7	(mod 9)	0	(mod 16)	4,32,40,68	$65+4^2=9^2$
2	(mod 9)	2	(mod 16)	2	7	(mod 9)		(mod 16)		
2	(mod 9)	3	(mod 16)	83	7	(mod 9)	1	(mod 16)	23,31,41,49	$371+23^2=30^2$
2	(mod 9)	4	(mod 16)	20	7	(mod 9)	0	(mod 16)	4,32,40,68	$20+4^2=6^2$
2	(mod 9)	5	(mod 16)	101	7	(mod 9)	4	(mod 16)	14,22,50,58	$101+50^2=51^2$
2	(mod 9)	6	(mod 16)	38	7	(mod 9)		(mod 16)		
2	(mod 9)	7	(mod 16)	119	7	(mod 9)	9	(mod 16)	5,13,59,67	$119+5^2=12^2$
2	(mod 9)	8	(mod 16)	56	7	(mod 9)	1	(mod 16)	5,13,59,67	$56+52=9^2$
2	(mod 9)	9	(mod 16)	137	7	(mod 9)	0	(mod 16)	4,32,40,68	$425+4^2=21^2$
2	(mod 9)	10	(mod 16)	74	7	(mod 9)		(mod 16)		
2	(mod 9)	11	(mod 16)	11	7	(mod 9)	9	(mod 16)	5,13,59,67	$155+13=18^2$
2	(mod 9)	12	(mod 16)	12	7	(mod 9)	4	(mod 16)		
2	(mod 9)	13	(mod 16)	13	7	(mod 9)	4	(mod 16)	14,22,50,58	$29+14^2=15^2$
2	(mod 9)	14	(mod 16)	14	7	(mod 9)		(mod 16)		
2	(mod 9)	15	(mod 16)	15	7	(mod 9)	1	(mod 16)	23,31,41,49	$335+31^2=36^2$

Table 12

Mod 9		Mod 16		Ex.	+ (mod 9)		(mod 16)		Possible	Example
3	(mod 9)	0	(mod 16)	192	1,4,7	(mod 9)	0,4	(mod 16)	2,8,22	$192+22^2=26^2$
3	(mod 9)	1	(mod 16)	273	1,4,7	(mod 9)	0,4	(mod 16)	4,16	$273+16^2=23^2$
3	(mod 9)	2	(mod 16)	66		(mod 9)		(mod 16)		
3	(mod 9)	3	(mod 16)	147	1,4,7	(mod 9)	1	(mod 16)	1,7,23	$147+23^2=26^2$
3	(mod 9)	4	(mod 16)	84	1,4,7	(mod 9)	0	(mod 16)	4, 20	$84+20^2=22^2$
3	(mod 9)	5	(mod 16)	165	1,4,7	(mod 9)	4	(mod 16)	2, 10	$165+26^2=29^2$
3	(mod 9)	6	(mod 16)	102		(mod 9)		(mod 16)		
3	(mod 9)	7	(mod 16)	39	1,4,7	(mod 9)	9	(mod 16)	5,19	$39+5^2=8^2$
3	(mod 9)	8	(mod 16)	1209	1,4,7	(mod 9)	9	(mod 16)	13,29	$120+13^2=17^2$
3	(mod 9)	9	(mod 16)	57	1,4,7	(mod 9)	0	(mod 16)	8,28	$57+28^2=29^2$
3	(mod 9)	10	(mod 16)	138		(mod 9)		(mod 16)		
3	(mod 9)	11	(mod 16)	75	1,4,7	(mod 9)	9	(mod 16)	5,37	$75+37^2=38^2$
3	(mod 9)	12	(mod 16)	156	1,4,7	(mod 9)	4	(mod 16)	10	$156+10^2=16^2$
3	(mod 9)	13	(mod 16)	93	1,4,7	(mod 9)	4	(mod 16)	14, 46	$93+14^2=17^2$
3	(mod 9)	14	(mod 16)	30		(mod 9)		(mod 16)		
3	(mod 9)	15	(mod 16)	111	1,4,7	(mod 9)	1	(mod 16)	17,55	$111+17^2=20^2$

Table 13

Mod 9		Mod 16		Ex.	+ (mod 9)		(mod 16)		Possible	Example
4	(mod 9)	0	(mod 16)	112	0	(mod 9)	0	(mod 16)	12,24,36,48,60	$112+12^2=16^2$
4	(mod 9)	1	(mod 16)	49	0	(mod 9)	0	(mod 16)	12,24,36,48,60	$481+12^2=25^2$
4	(mod 9)	2	(mod 16)	130		(mod 9)		(mod 16)		
4	(mod 9)	3	(mod 16)	67	0	(mod 9)	1	(mod 16)	9,15,33,39,57,63	$355+33^2=38^2$
4	(mod 9)	4	(mod 16)	4	0	(mod 9)	0	(mod 16)	12,24,36,48,60	$148+36^2=38^2$
4	(mod 9)	5	(mod 16)	85	0	(mod 9)	4	(mod 16)	6,18,30,42,54,66	$805+78^2=83^2$
4	(mod 9)	6	(mod 16)	22		(mod 9)		(mod 16)		
4	(mod 9)	7	(mod 16)	103	0	(mod 9)	9	(mod 16)	3,21,27,45,51,69	$247+3^2=16^2$
4	(mod 9)	8	(mod 16)	40	0	(mod 9)	1	(mod 16)	9,15,33,39,57,63	$184+21^2=25^2$
4	(mod 9)	9	(mod 16)	121	0	(mod 9)	0	(mod 16)	12,24,36,48,60	$265+24^2=29^2$
4	(mod 9)	10	(mod 16)	58		(mod 9)		(mod 16)		
4	(mod 9)	11	(mod 16)	139	0	(mod 9)	9	(mod 16)	3,21,27,45,51,69	$427+27^2=34^2$
4	(mod 9)	12	(mod 16)	76	0	(mod 9)	4	(mod 16)	6,18,30,42,54,66	$76+18^2=20^2$
4	(mod 9)	13	(mod 16)	13	0	(mod 9)	4	(mod 16)	9,15,33,39,57,63	$3-1+18^2=25^2$
4	(mod 9)	14	(mod 16)	94		(mod 9)		(mod 16)		
4	(mod 9)	15	(mod 16)	31	0	(mod 9)	1	(mod 16)	9,15,33,39,57,63	$319+9^2=20^2$

Table 14

Mod 9		Mod 16		Ex.	+ (mod 9)		(mod 16)		Possible	Example
5	(mod 9)	0	(mod 16)	32	4	(mod 9)	4	(mod 16)	2,34,38,70	$608+34^2=42^2$
5	(mod 9)	1	(mod 16)	113	4	(mod 9)	0	(mod 16)	16,20,52,56	$545+52^2=57^2$
5	(mod 9)	2	(mod 16)	50		(mod 9)		(mod 16)		
5	(mod 9)	3	(mod 16)	131	4	(mod 9)	1	(mod 16)	7,25,47,65	$275+25^2=30^2$
5	(mod 9)	4	(mod 16)	68	4	(mod 9)	0	(mod 16)	16,20,52,56	$68+16^2=18^2$
5	(mod 9)	5	(mod 16)	5	4	(mod 9)	4	(mod 16)	2,34,38,70	$437+2^2=21^2$
5	(mod 9)	6	(mod 16)	86		(mod 9)		(mod 16)		
5	(mod 9)	7	(mod 16)	23	4	(mod 9)	9	(mod 16)	11,29,43,61	$203+11^2=18^2$
5	(mod 9)	8	(mod 16)	104	4	(mod 9)	1	(mod 16)	7,25,47,65	$140+34^2=36^2$
5	(mod 9)	9	(mod 16)	41	4	(mod 9)	0	(mod 16)	16,20,52,56	$185+16^2=21¡$
5	(mod 9)	10	(mod 16)	122		(mod 9)		(mod 16)		
5	(mod 9)	11	(mod 16)	59	4	(mod 9)	9	(mod 16)	11,29,43,61	$203+11^2=18^2$
5	(mod 9)	12	(mod 16)	140	4	(mod 9)	4	(mod 16)	2,34,38,70	$284+70^2=72^2$
5	(mod 9)	13	(mod 16)	77	4	(mod 9)	4	(mod 16)	2,34,38,70	$77+38^2=39^2$
5	(mod 9)	14	(mod 16)	14		(mod 9)		(mod 16)		
5	(mod 9)	15	(mod 16)	95	4	(mod 9)	1	(mod 16)	7,25,47,65	$527+7^2=24^2$

Table 15

Mod 9		Mod 16		Ex.	+ (mod 9)		(mod 16)		Possible	Example
6	(mod 9)	0	(mod 16)	96	1,4,7	(mod 9)	4	(mod 16)	2,10,23	$96+23^2=25^2$
6	(mod 9)	1	(mod 16)	177	1,4,7	(mod 9)	0	(mod 16)	16,28	$177+28^2=31^2$
6	(mod 9)	2	(mod 16)	114		(mod 9)		(mod 16)		
6	(mod 9)	3	(mod 16)	51	1,4,7	(mod 9)	1	(mod 16)	7,25	$51+25^2=26^2$
6	(mod 9)	4	(mod 16)	132	1,4,7	(mod 9)	0	(mod 16)	4,12	$132+32^2=34^2$
6	(mod 9)	5	(mod 16)	69	0	(mod 9)	4	(mod 16)	10,34	$69+34^2=35^2$
6	(mod 9)	6	(mod 16)	6		(mod 9)		(mod 16)		
6	(mod 9)	7	(mod 16)	87	0	(mod 9)	0	(mod 16)	13,43	$87+13^2=16^2$
6	(mod 9)	8	(mod 16)	168	0	(mod 9)	1,9	(mod 16)	1,11,41	$168+1^2=13^2$
6	(mod 9)	9	(mod 16)	105	0	(mod 9)	0	(mod 16)	8,52	$105+8^2=13^2$
6	(mod 9)	10	(mod 16)	186		(mod 9)		(mod 16)		
6	(mod 9)	11	(mod 16)	123	1,4,7	(mod 9)	9	(mod 16)	19,61	$123+61^2=62^2$
6	(mod 9)	12	(mod 16)	60	1,4,7	(mod 9)	4	(mod 16)	14, 50	$60+14^2=16^2$
6	(mod 9)	13	(mod 16)	141	1,4,7	(mod 9)	4	(mod 16)	7,73	$147+73^2=74^2$
6	(mod 9)	14	(mod 16)	78		(mod 9)		(mod 16)		
6	(mod 9)	15	(mod 16)	159	1,4,7	(mod 9)	1	(mod 16)	7,25	$159+79^2=80^2$

Table 16

Mod 9		Mod 16		Ex.	+ (mod 9)		(mod 16)		Possible	Example
7	(mod 9)	0	(mod 16)	16	0	(mod 9)	0	(mod 16)	12,24,36,48,60	$304+36^2=40^2$
7	(mod 9)	1	(mod 16)	97	0	(mod 9)	0	(mod 16)	12,24,36,48,60	$385+12^2=23^2$
7	(mod 9)	2	(mod 16)	34	0	(mod 9)		(mod 16)		
7	(mod 9)	3	(mod 16)	115	0	(mod 9)	1	(mod 16)	9,15,33,39,57,63	$259+15^2=22^2$
7	(mod 9)	4	(mod 16)	52	0	(mod 9)		(mod 16)		
7	(mod 9)	5	(mod 16)	133	0	(mod 9)	4	(mod 16)	6,18,30,42,54,66	$565+54^2=59^2$
7	(mod 9)	6	(mod 16)	70	0	(mod 9)		(mod 16)		
7	(mod 9)	7	(mod 16)	7	0	(mod 9)	9	(mod 16)	3,21,27,45,51,69	$295+27^2=32^2$
7	(mod 9)	8	(mod 16)	88	0	(mod 9)		(mod 16)		
7	(mod 9)	9	(mod 16)	25	0	(mod 9)	0	(mod 16)	12,24,36,48,60	$25+12^2=13^2$
7	(mod 9)	10	(mod 16)	106	0	(mod 9)		(mod 16)		
7	(mod 9)	11	(mod 16)	43	0	(mod 9)	9	(mod 16)	3,21,27,45,51,69	$475+3^2=22^2$
7	(mod 9)	12	(mod 16)	124	0	(mod 9)		(mod 16)		
7	(mod 9)	13	(mod 16)	61	0	(mod 9)	4	(mod 16)	6,18,30,42,54,66	$205+18^2=23^2$
7	(mod 9)	14	(mod 16)	142	0	(mod 9)		(mod 16)		
7	(mod 9)	15	(mod 16)	79	0	(mod 9)	1	(mod 16)	9,15,33,39,57,63	$511+33^2=40^2$

Table 17

Mod 9		Mod 16		Ex.	+ (mod 9)		(mod 16)		Possible	Example
8	(mod 9)	0	(mod 16)	80	1	(mod 9)	1	(mod 16)	8,28,44,64	$80+8^2=12^2$
8	(mod 9)	1	(mod 16)	17	1	(mod 9)		(mod 16)	8,28,44,64	$161+8^2=15^2$
8	(mod 9)	2	(mod 16)	98	1	(mod 9)		(mod 16)		
8	(mod 9)	3	(mod 16)	35	1	(mod 9)	1	(mod 16)	1,17,55,71	$611+17^2=30^2$
8	(mod 9)	4	(mod 16)	116	1	(mod 9)		(mod 16)		
8	(mod 9)	5	(mod 16)	53	1	(mod 9)	4	(mod 16)	10,26,46,62	$341+10^2=21^2$
8	(mod 9)	6	(mod 16)	134	1	(mod 9)		(mod 16)		
8	(mod 9)	7	(mod 16)	71	1	(mod 9)	9	(mod 16)	19,35,37,53	$215+19^2=24^2$
8	(mod 9)	8	(mod 16)	8	1	(mod 9)		(mod 16)		
8	(mod 9)	9	(mod 16)	89	1	(mod 9)		(mod 16)	8,28,44,64	$377+8^2=21^2$
8	(mod 9)	10	(mod 16)	26	1	(mod 9)		(mod 16)		
8	(mod 9)	11	(mod 16)	107	1	(mod 9)	9	(mod 16)	19,35,37,53	$395+37^2=42^2$
8	(mod 9)	12	(mod 16)	44	1	(mod 9)		(mod 16)		
8	(mod 9)	13	(mod 16)	125	1	(mod 9)	4	(mod 16)	10,26,46,62	$413+26^2=33^2$
8	(mod 9)	14	(mod 16)	62	1	(mod 9)		(mod 16)		
8	(mod 9)	15	(mod 16)	143	1	(mod 9)	1	(mod 16)	1,17,55,71	$1007+17^2=36^2$

Table 18

Mod 9		Mod 16		Ex.	+ (mod 9)		(mod 16)		Possible	Example
0	(mod 9)	0	(mod 16)	144	1,4,7	(mod 9)	1,4,9	(mod 16)		$144+35^2=37^2$
0	(mod 9)	1	(mod 16)	81	1,4,7	(mod 9)	0	(mod 16)	12,4	$81+12^2=15^2$
0	(mod 9)	2	(mod 16)	18		(mod 9)		(mod 16)		
0	(mod 9)	3	(mod 16)	99	1,4,7	(mod 9)	1	(mod 16)	1,15,49	$99+49^2=50^2$
0	(mod 9)	4	(mod 16)	180	1,4,7	(mod 9)	0	(mod 16)	8,44	$180+44^2=46^2$
0	(mod 9)	5	(mod 16)	117	0	(mod 9)	4	(mod 16)	2,18	$117+2^2=11^2$
0	(mod 9)	6	(mod 16)	54		(mod 9)		(mod 16)		
0	(mod 9)	7	(mod 16)	135	0	(mod 9)	9	(mod 16)	alle 0 (mod 3)	$135+3^2=12^2$
0	(mod 9)	8	(mod 16)	72	0	(mod 9)		(mod 16)	9,15,33,39,57,63	$72+3^2=9^2$
0	(mod 9)	9	(mod 16)	153	0	(mod 9)	1,4,9	(mod 16)	4,24,76	$153+4^2=13^2$
0	(mod 9)	10	(mod 16)	90		(mod 9)		(mod 16)		
0	(mod 9)	11	(mod 16)	27	1,4,7	(mod 9)	1,4,9	(mod 16)	3,13	$27+13^2=14^2$
0	(mod 9)	12	(mod 16)	108	1,4,7	(mod 9)	4	(mod 16)		$28+6^2=8^2$
0	(mod 9)	13	(mod 16)	45	1,4,7	(mod 9)	4	(mod 16)	6,22	$45+6^2=9^2$
0	(mod 9)	14	(mod 16)	126		(mod 9)		(mod 16)		
0	(mod 9)	15	(mod 16)	63	1,4,7	(mod 9)	1	(mod 16)	9,31	$415+39^2=44^2$

Table 19

Mod 9		Mod 16		Ex.	+ (mod 9)		(mod 16)		Possible	Example
1	(mod 9)	0	(mod 16)	64	0	(mod 9)	all	(mod 16)	all 0 (mod 3)	$64+6^2=10^2$
1	(mod 9)	1	(mod 16)	1	0	(mod 9)	0	(mod 16)	6,12,18,24,36,48	$145+12^2=17^2$
1	(mod 9)	2	(mod 16)	82	0	(mod 9)		(mod 16)		
1	(mod 9)	3	(mod 16)	19	0	(mod 9)	1	(mod 16)	9,15,33,39,57,63	$451+15^2=26^2$
1	(mod 9)	4	(mod 16)	100	0	(mod 9)	0	(mod 16)	12,24,36,48,60	$100+24^2=26^2$
1	(mod 9)	5	(mod 16)	37	0	(mod 9)	4	(mod 16)	6,18,30,42,54,66	$325+6^2=19^2$
1	(mod 9)	6	(mod 16)	118	0	(mod 9)		(mod 16)		
1	(mod 9)	7	(mod 16)	55	0	(mod 9)	9	(mod 16)	alle 0 (mod 3)	$775+3^2=28^2$
1	(mod 9)	8	(mod 16)	136	0	(mod 9)		(mod 16)	9,15,33,39,57,63	$136+33^2=35^2$
1	(mod 9)	9	(mod 16)	73	0	(mod 9)	1,4,9	(mod 16)	12,24,36,48,60	$217+12^2=19^2$
1	(mod 9)	10	(mod 16)	10	0	(mod 9)		(mod 16)		
1	(mod 9)	11	(mod 16)	91	0	(mod 9)	1,4,9	(mod 16)	0	$91+3^2=10^2$
1	(mod 9)	12	(mod 16)	28	0	(mod 9)	4	(mod 16)	6,12,18,24,36,48	$28+6^2=8^2$
1	(mod 9)	13	(mod 16)	109	0	(mod 9)	4	(mod 16)	6,12,18,24,36,48	$253+6^2=17^2$
1	(mod 9)	14	(mod 16)	46	0	(mod 9)		(mod 16)		
1	(mod 9)	15	(mod 16)	127	0	(mod 9)	1	(mod 16)	9,15,33,39,57,63	$415+39^2=44^2$

Table 20

Mod 9		Mod 16		Ex.	+ (mod 9)		(mod 16)		Possible	Example
2	(mod 9)	0	(mod 16)	128	7	(mod 9)	0	(mod 16)	4,32,40,68	$128+14^2=18^2$
2	(mod 9)	1	(mod 16)	65	7	(mod 9)	0	(mod 16)	4,32,40,68	$65+4^2=9^2$
2	(mod 9)	2	(mod 16)	2	7	(mod 9)		(mod 16)		
2	(mod 9)	3	(mod 16)	83	7	(mod 9)	1	(mod 16)	23,31,41,49	$371+23^2=30^2$
2	(mod 9)	4	(mod 16)	20	7	(mod 9)	0	(mod 16)	4,32,40,68	$20+4^2=6^2$
2	(mod 9)	5	(mod 16)	101	7	(mod 9)	4	(mod 16)	14,22,50,58	$101+50^2=51^2$
2	(mod 9)	6	(mod 16)	38	7	(mod 9)		(mod 16)		
2	(mod 9)	7	(mod 16)	119	7	(mod 9)	9	(mod 16)	5,13,59,67	$119+5^2=12^2$
2	(mod 9)	8	(mod 16)	56	7	(mod 9)	1	(mod 16)	5,13,59,67	$56+52=9^2$
2	(mod 9)	9	(mod 16)	137	7	(mod 9)	0	(mod 16)	4,32,40,68	$425+4^2=21^2$
2	(mod 9)	10	(mod 16)	74	7	(mod 9)		(mod 16)		
2	(mod 9)	11	(mod 16)	11	7	(mod 9)	9	(mod 16)	5,13,59,67	$155+13=18^2$
2	(mod 9)	12	(mod 16)	12	7	(mod 9)	4	(mod 16)		
2	(mod 9)	13	(mod 16)	13	7	(mod 9)	4	(mod 16)	14,22,50,58	$29+14^2=15^2$
2	(mod 9)	14	(mod 16)	14	7	(mod 9)		(mod 16)		
2	(mod 9)	15	(mod 16)	15	7	(mod 9)	1	(mod 16)	23,31,41,49	$335+31^2=36^2$

Table 21

Mod 9		Mod 16		Ex.	To complete with + (mod 9)		(mod 16)		Possible	Example
3	(mod 9)	0	(mod 16)	192	1,4,7	(mod 9)	0,4	(mod 16)	2,8,22	$192+22^2=26^2$
3	(mod 9)	1	(mod 16)	273	1,4,7	(mod 9)	0,4	(mod 16)	4,16	$273+16^2=23^2$
3	(mod 9)	2	(mod 16)	66		(mod 9)		(mod 16)		
3	(mod 9)	3	(mod 16)	147	1,4,7	(mod 9)	1	(mod 16)	1,7,23	$147+23^2=26^2$
3	(mod 9)	4	(mod 16)	84	1,4,7	(mod 9)	0	(mod 16)	4, 20	$84+20^2=22^2$
3	(mod 9)	5	(mod 16)	165	1,4,7	(mod 9)	4	(mod 16)	2, 10	$165+26^2=29^2$
3	(mod 9)	6	(mod 16)	102		(mod 9)		(mod 16)		
3	(mod 9)	7	(mod 16)	39	1,4,7	(mod 9)	9	(mod 16)	5,19	$39+5^2=8^2$
3	(mod 9)	8	(mod 16)	120	1,4,7	(mod 9)	9	(mod 16)	13,29	$120+13^2=17^2$
3	(mod 9)	9	(mod 16)	57	1,4,7	(mod 9)	0	(mod 16)	8,28	$57+28^2=29^2$
3	(mod 9)	10	(mod 16)	138		(mod 9)		(mod 16)		
3	(mod 9)	11	(mod 16)	75	1,4,7	(mod 9)	9	(mod 16)	5,37	$75+37^2=38^2$
3	(mod 9)	12	(mod 16)	156	1,4,7	(mod 9)	4	(mod 16)	10	$156+10^2=16^2$
3	(mod 9)	13	(mod 16)	93	1,4,7	(mod 9)	4	(mod 16)	14, 46	$93+14^2=17^2$
3	(mod 9)	14	(mod 16)	30		(mod 9)		(mod 16)		
3	(mod 9)	15	(mod 16)	111	1,4,7	(mod 9)	1	(mod 16)	17,55	$111+17^2=20^2$

Table 22

Mod 9		Mod 16		Ex.	To complete with + (mod 9)		(mod 16)		Possible	Example
4	(mod 9)	0	(mod 16)	112	0	(mod 9)	0	(mod 16)	12,24,36,48,60	$112+12^2=16^2$
4	(mod 9)	1	(mod 16)	49	0	(mod 9)	0	(mod 16)	12,24,36,48,60	$481+12^2=25^2$
4	(mod 9)	2	(mod 16)	130		(mod 9)		(mod 16)		
4	(mod 9)	3	(mod 16)	67	0	(mod 9)	1	(mod 16)	9,15,33,39,57,63	$355+33^2=38^2$
4	(mod 9)	4	(mod 16)	4	0	(mod 9)	0	(mod 16)	12,24,36,48,60	$148+36^2=38^2$
4	(mod 9)	5	(mod 16)	85	0	(mod 9)	4	(mod 16)	6,18,30,42,54,66	$805+78^2=83^2$
4	(mod 9)	6	(mod 16)	22		(mod 9)		(mod 16)		
4	(mod 9)	7	(mod 16)	103	0	(mod 9)	9	(mod 16)	3,21,27,45,51,69	$247+3^2=16^2$
4	(mod 9)	8	(mod 16)	40	0	(mod 9)	1	(mod 16)	9,15,33,39,57,63	$184+21^2=25^2$
4	(mod 9)	9	(mod 16)	121	0	(mod 9)	0	(mod 16)	12,24,36,48,60	$265+24^2=29^2$
4	(mod 9)	10	(mod 16)	58		(mod 9)		(mod 16)		
4	(mod 9)	11	(mod 16)	139	0	(mod 9)	9	(mod 16)	3,21,27,45,51,69	$427+27^2=34^2$
4	(mod 9)	12	(mod 16)	76	0	(mod 9)	4	(mod 16)	6,18,30,42,54,66	$76+18^2=20^2$
4	(mod 9)	13	(mod 16)	13	0	(mod 9)	4	(mod 16)	9,15,33,39,57,63	$3-1+18^2=25^2$
4	(mod 9)	14	(mod 16)	94		(mod 9)		(mod 16)		
4	(mod 9)	15	(mod 16)	31	0	(mod 9)	1	(mod 16)	9,15,33,39,57,63	$319+9^2=20^2$

Table 23

Mod 9		Mod 16		Ex.	To complete with + (mod 9)		(mod 16)		Possible	Example
5	(mod 9)	0	(mod 16)	32	4	(mod 9)	4	(mod 16)	2,34,38,70	$608+34^2=42^2$
5	(mod 9)	1	(mod 16)	113	4	(mod 9)	0	(mod 16)	16,20,52,56	$545+52^2=57^2$
5	(mod 9)	2	(mod 16)	50		(mod 9)		(mod 16)		
5	(mod 9)	3	(mod 16)	131	4	(mod 9)	1	(mod 16)	7,25,47,65	$275+25^2=30^2$
5	(mod 9)	4	(mod 16)	68	4	(mod 9)	0	(mod 16)	16,20,52,56	$68+16^2=18^2$
5	(mod 9)	5	(mod 16)	5	4	(mod 9)	4	(mod 16)	2,34,38,70	$437+2^2=21^2$
5	(mod 9)	6	(mod 16)	86		(mod 9)		(mod 16)		
5	(mod 9)	7	(mod 16)	23	4	(mod 9)	9	(mod 16)	11,29,43,61	$203+11^2=18^2$
5	(mod 9)	8	(mod 16)	104	4	(mod 9)	1	(mod 16)	7,25,47,65	$140+34^2=36^2$
5	(mod 9)	9	(mod 16)	41	4	(mod 9)	0	(mod 16)	16,20,52,56	$185+16^2=21¡$
5	(mod 9)	10	(mod 16)	122		(mod 9)		(mod 16)		
5	(mod 9)	11	(mod 16)	59	4	(mod 9)	9	(mod 16)	11,29,43,61	$203+11^2=18^2$
5	(mod 9)	12	(mod 16)	140	4	(mod 9)	4	(mod 16)	2,34,38,70	$284+70^2=72^2$
5	(mod 9)	13	(mod 16)	77	4	(mod 9)	4	(mod 16)	2,34,38,70	$77+38^2=39^2$
5	(mod 9)	14	(mod 16)	14		(mod 9)		(mod 16)		
5	(mod 9)	15	(mod 16)	95	4	(mod 9)	1	(mod 16)	7,25,47,65	$527+7^2=24^2$

Table 24

Mod 9		Mod 16		Ex.	+ (mod 9)		(mod 16)		Possible	Example
6	(mod 9)	0	(mod 16)	96	1,4,7	(mod 9)	4	(mod 16)	2,10,23	$96+23^2=25^2$
6	(mod 9)	1	(mod 16)	177	1,4,7	(mod 9)	0	(mod 16)	16,28	$177+28^2=31^2$
6	(mod 9)	2	(mod 16)	114		(mod 9)		(mod 16)		
6	(mod 9)	3	(mod 16)	51	1,4,7	(mod 9)	1	(mod 16)	7,25	$51+25^2=26^2$
6	(mod 9)	4	(mod 16)	132	1,4,7	(mod 9)	0	(mod 16)	4,12	$132+32^2=34^2$
6	(mod 9)	5	(mod 16)	69	0	(mod 9)	4	(mod 16)	10,34	$69+34^2=35^2$
6	(mod 9)	6	(mod 16)	6		(mod 9)		(mod 16)		
6	(mod 9)	7	(mod 16)	87	0	(mod 9)	0	(mod 16)	13,43	$87+13^2=16^2$
6	(mod 9)	8	(mod 16)	168	0	(mod 9)	1,9	(mod 16)	1,11,41	$168+1^2=13^2$
6	(mod 9)	9	(mod 16)	105	0	(mod 9)	0	(mod 16)	8,52	$105+8^2=132$
6	(mod 9)	10	(mod 16)	186		(mod 9)		(mod 16)		
6	(mod 9)	11	(mod 16)	123	1,4,7	(mod 9)	9	(mod 16)	19,61	$123+61^2=62^2$
6	(mod 9)	12	(mod 16)	60	1,4,7	(mod 9)	4	(mod 16)	14, 50	$60+14^2=16^2$
6	(mod 9)	13	(mod 16)	141	1,4,7	(mod 9)	4	(mod 16)	7,73	$147+73^2=74^2$
6	(mod 9)	14	(mod 16)	78		(mod 9)		(mod 16)		
6	(mod 9)	15	(mod 16)	159	1,4,7	(mod 9)	1	(mod 16)	7,25	$159+79^2=80^2$

Table 25

Mod 9		Mod 16		Ex.	+ (mod 9)		(mod 16)		Possible	Example
7	(mod 9)	0	(mod 16)	16	0	(mod 9)	0	(mod 16)	12,24,36,48,60	$304+36^2=40^2$
7	(mod 9)	1	(mod 16)	97	0	(mod 9)	0	(mod 16)	12,24,36,48,60	$385+12^2=23^2$
7	(mod 9)	2	(mod 16)	34	0	(mod 9)		(mod 16)		
7	(mod 9)	3	(mod 16)	115	0	(mod 9)	1	(mod 16)	9,15,33,39,57,63	$259+15^2=22^2$
7	(mod 9)	4	(mod 16)	52	0	(mod 9)		(mod 16)		
7	(mod 9)	5	(mod 16)	133	0	(mod 9)	4	(mod 16)	6,18,30,42,54,66	$565+54^2=59^2$
7	(mod 9)	6	(mod 16)	70	0	(mod 9)		(mod 16)		
7	(mod 9)	7	(mod 16)	7	0	(mod 9)	9	(mod 16)	3,21,27,45,51,69	$295+27^2=32^2$
7	(mod 9)	8	(mod 16)	88	0	(mod 9)		(mod 16)		
7	(mod 9)	9	(mod 16)	25	0	(mod 9)	0	(mod 16)	12,24,36,48,60	$25+12^2=13^2$
7	(mod 9)	10	(mod 16)	106	0	(mod 9)		(mod 16)		
7	(mod 9)	11	(mod 16)	43	0	(mod 9)	9	(mod 16)	3,21,27,45,51,69	$475+3^2=22^2$
7	(mod 9)	12	(mod 16)	124	0	(mod 9)		(mod 16)		
7	(mod 9)	13	(mod 16)	61	0	(mod 9)	4	(mod 16)	6,18,30,42,54,66	$205+18^2=23^2$
7	(mod 9)	14	(mod 16)	142	0	(mod 9)		(mod 16)		
7	(mod 9)	15	(mod 16)	79	0	(mod 9)	1	(mod 16)	9,15,33,39,57,63	$511+33^2=40^2$

Table 26

Mod 9		Mod 16		Ex.	+ (mod 9)		(mod 16)		Possible	Example
8	(mod 9)	0	(mod 16)	80	1	(mod 9)	1	(mod 16)	8,28,44,64	$80+8^2=12^2$
8	(mod 9)	1	(mod 16)	17	1	(mod 9)		(mod 16)	8,28,44,64	$161+8^2=15^2$
8	(mod 9)	2	(mod 16)	98	1	(mod 9)		(mod 16)		
8	(mod 9)	3	(mod 16)	35	1	(mod 9)	1	(mod 16)	1,17,55,71	$611+17^2=30^2$
8	(mod 9)	4	(mod 16)	116	1	(mod 9)		(mod 16)		
8	(mod 9)	5	(mod 16)	53	1	(mod 9)	4	(mod 16)	10,26,46,62	$341+10^2=21^2$
8	(mod 9)	6	(mod 16)	134	1	(mod 9)		(mod 16)		
8	(mod 9)	7	(mod 16)	71	1	(mod 9)	9	(mod 16)	19,35,37,53	$215+19^2=24^2$
8	(mod 9)	8	(mod 16)	8	1	(mod 9)		(mod 16)		
8	(mod 9)	9	(mod 16)	89	1	(mod 9)		(mod 16)	8,28,44,64	$377+8^2=21^2$
8	(mod 9)	10	(mod 16)	26	1	(mod 9)		(mod 16)		
8	(mod 9)	11	(mod 16)	107	1	(mod 9)	9	(mod 16)	19,35,37,53	$395+37^2=42^2$
8	(mod 9)	12	(mod 16)	44	1	(mod 9)		(mod 16)		
8	(mod 9)	13	(mod 16)	125	1	(mod 9)	4	(mod 16)	10,26,46,62	$413+26^2=33^2$
8	(mod 9)	14	(mod 16)	62	1	(mod 9)		(mod 16)		
8	(mod 9)	15	(mod 16)	143	1	(mod 9)	1	(mod 16)	1,17,55,71	$1007+17^2=36^2$

11 Structure in the cubes

Table 1

Basic Number	Cubed
1	1
2	8
3	27
4	64
5	125
6	216
7	343
8	512
9	729

Looking at table 1 we find some things which strike us.

- 1, 4, 5, 6 and 9 remain unchanged
- 2, 3, 7 and 8 are interchanged
- The number of digits of the answer varies from 1 to 3

When calculating to the third power, the number of figures of the result will be basic number times 3, eventually minus 1 or minus 2. So for a three figure number the cubes can have from 7 up to 9 figures. This is the quantitative approach. When we look for the final figures and study them then we speak of the qualitative approach.

Table 2

B.N.	Cube	B.N.	Cube	B.N.	Cube	B.N.	Cube	B.N.	Cube
1	01	2	8	3	27	4	64	5	125
11	1331	12	1728	13	22197	14	2744	15	3375
21	9261	22	10648	23	12167	24	13824	25	15625
31	29791	32	32768	33	35937	34	39304	35	42875
41	68921	42	74088	43	79507	44	85184	45	91125
51	132651	52	140608	53	148877	54	157464	55	166375
61	226981	62	238328	63	250047	64	262144	65	274625
71	357911	72	373248	73	389017	74	405224	75	421875
81	531441	82	551368	83	571787	84	592704	85	614125
91	753571	92	778688	93	804357	94	830584	95	857375

B.N. means Basic Number. Cube means that the basic number is calculated to the third power.

Let's have a look on the column of the cubes ending with 1, and then especially on the tens of the numbers (the last digits as you see are all ones). We see 01, 31, 61, 91, 21, et cetera. It is clear: the tens increase by 3. Where does that come from? Mr. Isaac Newton found out; the method bears his name. The tens increase according to this form: 3× number of tens squared × the number of the "new" ten. So from 1 to 11 the tens increase with 3 × 1 = 3 and we find this confirmed through the whole column. So the cube of 71 will have a ten of $3 \times 1^2 \times 7$, which ends with 1. And indeed, we see that 71^3 = 3579<u>1</u>1.

Now for the column of the twos. The tens increase by two and obey the same formula. The increase of the tens is $3 \times 2^2 \times$ the number of tens. Per ten there is an increase of 2; see the column. For $22^3 = 3 \times 2^2 \times 2 = 4$, for 32^3 the tens are $3 \times 2^2 \times 3 = 6$, and indeed $32^3 = 327\underline{6}8$.

Now the threes. Increase of the tens per ten: $3 \times 3^2 = 1 = (2)7$; see the column. E.g. for 43^3 the tens increase with $3 \times 3^2 \times 4 = 8$ and indeed for 43^3 the tens are 2 + 8 = 0; the answer is 795<u>0</u>7.

The fours. For 34^3 the tens increase with $3 \times 4^2 =(4)$ 8. So 34^3 has a ten of 6 (4^3) + 3 × $4^2 \times 3 = (14)4 = 6 + 4 = (1)0$. And indeed $34^3 = 39304$.

The fives are special ones. Here we see an increase of 250 per 10. Later on more about this in "The cubic fives"; a document especially written about integer cubic roots of numbers ending in fives.

Now we see how the structure of the tens develops, we'll examine if the hundreds' increase goes in the same way. The increase of the hundreds is called a 'Jump'.

Table 3

1	1	Jump	2	8	Jump	3	27	Jump	4	64	Jump
101	1030301	300	102	1061208	1200	103	1092727	2700	104	1124864	4800
201	8120601		202	8242408		203	8365427		204	8489664	
301	27270901		302	27543608		303	27818127		304	28094464	
401	64481201		402	64964808		403	65450827		404	65939264	
501	125751501		502	126506008		503	127263527		504	128024064	
601	217081801		602	218167208		603	219256227		604	220348864	
701	344472101		702	345948408		703	347428927		704	348913664	
801	513922401		802	515849608		803	517781627		804	519718464	
901	731432701		902	733870808		903	736314327		904	738763264	

The formula for the jump of the hundreds is the same as for the tens. So for numbers x hundred 01 the jump per hundred is $3 \times 1^2 \times 100 = 300$. In the same way, the jump for x hundred 02 numbers is $3 \times 2^2 \times 100 = 1200$ per hundred. For 1473 the hundreds are 73^3 = 389017 + $3 \times 73^2 \times 1400 = 1800$, together 0817, so will end on 0817 as 1473^3 =

Behaving in tn the same way are the thousands. For 1073^3 the jump is $3 \times 1^2 \times 73^2 \times 1000 = (159)\ 87000$. And look: $1073^3 = 12353\ \mathbf{76017}$, from 89017 to 76017. The jump formula works perfectly, but one should not forget that if one wants to know the thousands of a cube, he should know the thousands of the basic number. E.g. jumping from 73^3 to 1873^3 with $18 \times 8700 = (15)6600$ we get indeed the hundreds: xxxx5617, but not the ten thousands. If we want to get the ten thousands, 873^3 should be the base. $873^3 = (6653)\ 38617$, $1873^3 = (65707)\ 25617$; indeed an increase of the thousands of 87000.

It is also possible to go from 73^3 to 1873^3, but then it is indispensable to calculate also $3 \times 1800^2 \times 73$ as the result of this has an impact on the ten thousands.

```
73³                           3 8 9 0 1 7
3 × 1800 × 73²              2 8 7 7 6 6 0 0
3 × 1800² × 73            7 0 9 5 6 0 0 0 0
1800³                   5 8 3 2 0 0 0 0 0 0
                        ─────────────────────
                        6 5 7 0 7 2 5 6 1 7
```

In the rule $3*1800^2 *73$ you can see that the 6 in the number 709560000 counts in the column of the ten thousands.

The fives behave differently. More about that now in the chapter "the cubic fives".

11.1 The cubic fives

As in the squares, the fives also play a special role in the cubes. In tournaments so far no question has been asked of the nature "integer cube roots". From this point of view an inquiry is made to find out if there are possibilities to recognise the basic number if the third power of it is given.

The first thing to do was to make a table. There I found a typical "jump", the difference between 2 succeeding numbers. This table is crucial. The jumps are marked in gray.
When we make a survey of the table we can simplify the jump numbers by dividing by 125. Then we get resp. 460, 140, 300, 140 and 460×125.

Table 4

05	15	25	35	45	
57500	17500	37500	17500	57500	Jumps per 100
460	**140**	**300**	**140**	**460**	× 125
55	65	75	85	95	
57500	17500	37500	17500	57500	Jumps per 100
460	**140**	**300**	**140**	**460**	× 125

Jumps per 100 means that e.g. 105^3 ends on $125 + 57500 = 57625$.
So 335^3 ends on $42875 + (3 \times 17500)\ 52500 = 95375$.

On studying table 4, the following is striking: We take 25 and 75 as the "centre numbers", and see that the "jump" per 100 equals 37500. This means that e.g. 325^3 ends with $15625 + 3 \times 37500 = (1)12500 = 28125$.

Furthermore, 1 "step" left and right from 25 and 75 we see that the jump is 17500 per 100. As 65^3 ends in (2)74625, and we want to know the last figures of 765^3, we add $7 \times 17500 = 22500$, so 765^3 ends in 97125. And indeed: $765^3 = (4476)97125$.

In the same way, 2 "steps" left and right from 25 and 75 we have a jump of 57500 per 100.

This is again the 'mirror' effect which we have already seen several times.
It is easy to remember: 25 and 75 have a jump of 37500. From each +10 and -10 we have a jump of 17500, +20 and -20 have a jump of 57500.

In the fives it is not sufficient to look at 3 or 4 digits; the 'advantage' of never changing fives has as a disadvantage that we have to extend our look to a bigger number of digits. The best thing to do is to take a rule and then see, 5 digits from the right, how the jumps in the cubic fives develop. Looking at table 6, on the next page, you'll see that indeed 115 and 4115 have the same last digits.

Particularly for integer cube roots of big numbers, mod 33 calculation is an indispensable tool. For these questions the ready knowledge of the cubes up to 100 is also indispensable.

Table 5 is to help you to get insight into the final figures of the cubes of the fives. For your ease, the numbers and their complements are put opposite to each other: 5^3 ends on 125, 395^3 ends on 9875. "Complement thinking" here is very important and helpful. Starting with 405 and further you'll get the same result with the four final figures.

So after 400 start a new series of numbers with the same four final figures. And after 4000 we have the same five final figures.

Table 5

5	125	61629875	395
15	3375	57066625	385
25	15625	52734375	375
35	42875	57066625	365
45	91125	44738875	355
55	166375	41063625	345
65	274625	37595375	335
75	421875	34328125	325
85	614125	31255875	315
95	857375	28372625	305
105	1157625	25672375	295
115	1520875	23149125	285
125	1953125	20796875	275
135	2460375	18609625	265
145	3048625	16581375	255
155	3723875	14706125	245
165	4492125	12977875	235
175	5359375	11390625	225
185	6331625	9938375	215
195	7414875	8615125	205

Table 6

BN	BN³	BN	BN³
15	03375	2015	8181353375
115	1520875	2115	9460870875
215	9938375	2215	10867288375
315	31255875	2315	12406605875
415	71473375	2415	14084823375
515	136590875	2515	15907940875
615	232608375	2615	17881958375
715	365525875	2715	20012875875
815	541343375	2815	22306693375
915	766060875	2915	24769410875
1015	1045678375	3015	27407028375
1115	1386195875	3115	30225545875
1215	1793613375	3215	33230963375
1315	2273930875	3315	36429280875
1415	2833148375	3415	39826498375
1515	3477265875	3515	43428615875
1615	4212283375	3615	47241633375
1715	5044200875	3715	51271550875
1815	5979018375	3815	55524368375
1915	7022735875	3915	60006085875
		4015	64722703375
		4115	69680220875

If we combine the jumps of table 4 with the contents of table 6 we can do a lot.

$$\sqrt[3]{324{,}951{,}171{,}875}$$

Which we subdivide as follows:

32 49 51 17 18 75

The question number has 12 figures, so the answer has 4. N(earest) L(ower) C(ube) = 314432 = 68³. So 68 are the first figures of the answer. Have a look at table 2: There we see that in the hundreds 0 mod 4 there is only one candidate which cube ends on 1875; 75. The final answer is 6875.

$$\sqrt[3]{5{,}182{,}207{,}647{,}625}$$

Subdivided as follows:

5 18 22 07 64 76 25, from right to left as mod 33: 25 + 76 + 64 + 07 + 22 + 18 + 5 = 19 (mod 33).

The question number has 13 figures, so the answer will have 5 figures. We start with 5182 to find the first two figures of the answer. N(earest) L(ower) C(ube) 4913 = 17^3. Last figures of the answer are 05 according to table 2, with an odd hundred. We now have 17 ? 05. As it is easier to work from right to left in groups of two figures we write: 1 7? 05. As the mod 33 of the question number = 19, the basic number is 13 (mod 33). 1 + 05 = 6 mod 33. To obtain 13 mod 33 we know 70 = 4 (mod 33), so we have to add 3 to get 73, which is 7 (mod 33). Write 73, and the final answer is 1 73 05.

$^3\sqrt{20{,}202{,}543{,}841{,}138{,}875}$

To be subdivided as follows: 2 + 02 + 02 + 54 + 38 + 41 + 13 + 88 + 75 = 18 (mod 33).

The question number has 17 figures, so the answer has 6. N(earest) L(ower) C(ube) = 19683 = 27^3, the first two figures of the answer are now known; 27. In table 2 you can see that the last four figures of the question number point to 355. For an even thousand there is also 755, for odd thousands the possibilities are 1155, 1555 and 1955.

The difference between 28^3, 21952 and 27^3, 19683 = 2269. Then: 20202 – 19683 = 519, which we divide by 2269, and come at ± 0.23. Our answer now is 27 23 55.

The mod 33 of the question number = 18. This means that the basic number will be 6 mod 33. 27 + 55 = 16 mod 33. To come from 16 to 6 mod 33 we have to add 23. Therefore our final answer will be 27 23 55.

Strictly thinking mod 33 56 and 89 are also possible. But the choice of 56 would lead to an odd hundred in the question number, the one we have here is an even one, so 56 drops out. 23 and 89 are so remote from each other that 89 is out of the race. Moreover; our division, which is a reliable approach, points to 0.23.

This find of Mr. Newton is really shrewd! What do we really do here? Firstly we determine the differences between the two cubes involved, 28^3 = 21952 – 27^3 = 19683 = 2269. This we call the "big difference". And we calculate the difference between the first five figures of the question number and the lowest cube number: 20202 – 19683 = 519. This we call the "small difference". By dividing the small difference by the big difference we get a reliable impression of the following figures of our answer. Nota bene: They give only an impression.

For elucidation we calculate 29^3 = 24389 and subtract 28^3 = 21952, the difference is 2437. It is clear that according as the numbers are getting bigger, the difference between two consecutive cubes is increasing progressively.

Now our answer is 27 23 55. The check: As the question number is 18 (mod 33), this means that the answer number will be 6 (mod 33). 27 + 23 + 55 = 105 = 6 (mod 33).

Strictly calculated via modulo 33, the answers could also be 23 56 05 or 23 89 05. But 56 would lead to an odd hundred in the question number, so this not correct. And 23 and 89 are that distant from each other that 89 drops out. In combination with the result of our division – the little difference divided by the big difference – we determine the answer to be 27 23 55.

$$\sqrt[3]{173{,}904{,}162{,}033{,}032{,}068{,}625}$$

The question has 21 figures, so the answer has seven figures. The subdivision is 1 + 73 + 90 + 41 + 62 + 03 + 30+ 32 + 06 + 86 + 25 = 449 = 20 (mod 33).

NLC = 55^3 = 166375, so the first figures of the answer are 55. According to table 2 the last figures of the answer are, in case of an even thousand, 145 + some 400. N(earest) H(igher) C(ube) 175616 = 56^3.

Difference between NHC and NLC = 175616 – 166375 = 9241, which we know as the big difference. The little difference is 173904 – 166375 = 7529. Now the division; little difference 7529 divided by the big difference 9241 = ± 0.81.

We now have as the answer 5 58 1? 45.

The mod 33 of the question number is calculated as 20, the basic number then is 26 mod 33. We add: 5 + 58 + 45 = 108 = 9 mod 33. 26 – 9 = 17 so the 1? = 17 and the final answer is 5 58 17 45.

11.2 Extracting integer cubic roots

For doing this kind of operation a ready knowledge of the cubes up to 100 is indispensable. For those who do not have this knowledge, please take a table with you, and then you'll see that with a not so big knowledge nevertheless the work is possible. Of course we'll work with numbers with more than 6 figures, as for these questions you do not need anything more than only the table. And of course we use the work from the chapter about the structure in the cubes.

$$\sqrt[3]{4{,}657{,}463}$$

4,657,463, seven figures. As 4096 < 4657 < 4913, the first figures of the answer are 16. As the last figure of the question number is a 3, the only figure which can follow will be 7, so our answer is 167.

$$\sqrt[3]{958{,}585{,}256}$$

As this number has nine figures we see immediately that the answer will have three figures. Even if we only know by heart the cubes up to 10 we'll see that:

- The first figure of the answer will be 9
- The last figure of the answer will be 6

And we use:

- Modulo calculation. The mod 9 of the question number is 8.
- the basic number can be 2, 5 or 8 (mod 9).
- The iteration of Newton. 958 − 729 (= 9^3) = 229, to be divided by 3 × 9^2 = 243. This is ± 0.94, so 8 is the correct choice.

$$\sqrt[3]{2{,}087{,}336{,}952}$$

As this number has ten figures, the answer will have 4 figures. Considering 1728 (12^3) < 2087 < 2197 (13^3) the first two figures of the answer are 12. As 2 is the last figure of the question number the last figure will be 8; the answer so far being 12 ? 8. The jump in the tens for 8 = 3 × 8^2 = (19) 2. 8^3 = 5 12. We start with 1 and go to 5, the jump is 4 in how many tens? Either 2 or 7 are the possibilities.

According to the Newton iteration we look at (13^3) = 2197 − (12^3) 1728 = 469 (the big difference). Then 2087 − 1728 = 359 (the small difference) ÷ 469 = 0.76, which means that our answer will be 1278.

If we want to work with mod 33 we do this: 20 + 87 + 33 + 69 + 52 = 228 = 30 (mod 33) which means that the answer will be 24 (mod 33). We have already 12, we know the last figure is 8, then the only number which fits is 78, and the final answer is (again) 1278.

Now we make a big step:

$$\sqrt[3]{588{,}290{,}886{,}482{,}671}$$

As this number has fifteen figures, the answer has five figures. As 571787 < 588290 < 592704 the first figures of the answer are 83. As the two last figures of the question number are 71 we find – in our memory or in our table – that the two last figures are 91, so that in the meantime we have as our provisional answer 83 ? 91. 91^3 = 753,571. The jump is 3 × 91^2 = (2484) 3, three per hundred. From 5(71) to 6(71) is a jump of 100, therefore the answer is 1(00) ÷ 3 Ξ 7, so our complete answer is 83791.

$$\sqrt[3]{13{,}610{,}149{,}100{,}785{,}216}$$

Which we think of as 1 + 36 + 10 + 14 + 91 + 00 + 78 + 52 + 16, seventeen figures, so the answer has six figures, the mod 11 = 1. Then our final answer will be 1 (mod 11) too.

As 12167< 13610< 13824, the answer will have 23 as the two first figures.

Looking at the last two figures we see that the last two figures of the answer are either 06 or 56.

This is the moment to introduce mod 16 calculation. If we take 6 in combination with an even ten, so 06, 26, 46 et cetera the number is only divisible by 2 and not by 4. Such a number calculated to the third power results in a number which is divisible by 8 but not by 16. As 5216 is divisible by 16, the 06 drops out. What we have now is 23 ?? 56.

Now Newton: 13824 − 12167 = 1657. 13610 − 12167 = 1443. 1443 ÷ 1657 = ± 0.87.

Next modulo 11 calculation: 23 + 56 = 79 = 2 (mod 11). And 2 + 87 [= 10 (mod 11)] means that 87 is the right number, our final answer then is 23 87 56.

$$\sqrt[3]{157{,}343{,}395{,}883{,}961{,}837{,}061}$$

Which we think as 157343 39588396183 7061. These are 21 figures, which means that our answer will have 7 figures.

According to the table we find that as 148877 < 157343 < 157464 that the first figures of our answer will be 53 and that the final answer will not be far from 5,400,000.

The last figures 61 mean that the last figures of our answer will be 21. Our answer so far is 53 xxx 21. 21^3 = 9261 and the jump in the hundreds is 3 × 21^2 × 100 = (mod 13) 2300, so 23 per hundred.

The question number ends on 7061 and 21^3 = 9261. The difference in hundreds is 7800. Now comes the question of how many steps of 23(00) do we need to arrive on 78, so 78 ÷ 23 ≡ 86, as 86 × 23 = (19) 78. The answer so far is 53 ? 86 21.

The missing figure can be found by working with the Newton iteration: 157343 − 148877 (53^3) = 8466 ÷ 8427 = 1 + something. This is not strange, as in fact we have to take 53 + something to subtract. So we take 9 and now have the complete answer 5 39 86 21.

Otherwise we can apply modulo 11. Working in groups of two figures 1 57 34 33 95 88 39 61 83 70 61 from right to left have 6 + 4 + 6 + 6 + 6 + 0 + 7 + 0 + 1 + 2 + 1. Let's be clever: 6 + 4 + 6 + 6 = 0; 7 + 1 + 2 + 1 = 0, remaining is 6. Some number $?^3$ = 6 (mod 11). According to the table we find this number is 8. So our basic number has to be 8 (mod 11).

86 + 21 = 8 (mod 11). 53? has to be 0 (mod 11). This can only be 539, by which means our answer comes to 5 39 86 21.

12 Finding prime numbers

In fact this activity is a kind of factorisation. Generally the numbers offered in the competitions to be factorised are composed by a lot of prime numbers, and their value does not exceed 100.

This chapter is written in memory of the German Mental Calculation prodigy Zacharias Dase, who lived from 1824–1861. He was presented with a nine figure number – 278,353,657 – which is composed of three 3-digit prime numbers, which are not consecutive. In the literature we find that Dase succeeded in finding the prime factors of this number in 29 minutes, without using paper or making intermediate calculations.
In literature is no mention of another calculator who repeated this performance, let alone in less time.

In a message the English calculation prodigy Dr. Robert Fountain invited me to try this, as I after all being "King of the Primes" were the only one world wide who could eventually do this. I accepted the challenge.

The term "finding prime numbers" generally means that the task concerns bigger numbers. In this chapter, to start with we explain the question numbers having six figures and which are composed of three two digit prime numbers.

After a chat with another MC prodigy, Jan van Koningsveld, the following technique was applied:

- Take the nearest higher prime number after the first figures of the question number
- Calculate the nearest round cube root to that of the complete question number
- Between these two numbers lies somewhere one of the three primes we are looking for
- Start with dividing the QN by primes between the numbers mentioned, as long as one has an integer division, so there is no remainder
- Next find the other two primes according to the method which will be explained

To gain confidence with this kind of question we start with a six digit number.

Procedure:

The question number is 283531. Divide the number roughly by the highest 2 digit prime (97), the answer is 29+. So 29 is our lowest candidate; 29 × 97 = 2813. Then take the rounded cube root of the question number, which is 65+, and the next lower prime number = 61.

Table 1

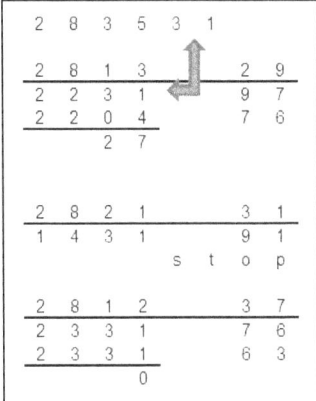

We now know that from the three two digit prime numbers we are looking for, anyhow one lies between 29 and 61. Which one is in fact a guess – Russian Roulette – there is no algorithm available with which we can pinpoint exactly the number we are so desperately looking for. Fortunately there are no lives at stake. There is no other method than "roughly dividing" until the moment we have an integer solution, i.e. a division without a remainder. The arrow indicates the appending of the last digits of the question number.

Ok, off we go!!

- The first division by 29 results in 97 times to start with, and we subtract the result 2813 from 2835 and get a remainder of 22
- 22 × 100 = 2200 + and append 31 to get 2231
- We now have 2231 which divided by 29 results in 76 times and a remainder of 27
- As 27 is not divisible by 29 this is not the one we want.
- Then we divide the question number by the next candidate, 31
- The division by 31 results in 91 times to start with and we subtract the result 2821 from 2835 and get a remainder of 14 and then append 31 to get 1431.
- We now have 1435 ÷ 31 = 46 r 9, so neither is 31 the number we are looking for.
- The third division, by 37, results in 76 times to start with and we subtract the result 2812 from 2835 and get a remainder of 23; then we append 31 to get 2331.
- 2331 divided by 37 results in 63 times and a zero remainder, answer 7663.
- This means we have now one of the three numbers we are looking for.

Based on the nature of the question – find the three two digit prime numbers which compose the question number 283531 – we now conclude that 7663 is the product of two 2 digit prime numbers, which we have to find.

These can be found as follows:

- 7663 = 4 (mod 9) and 15 (mod 16).
- Based on (a + b) × (a – b) we look for a smaller square to complete 7663 to another – bigger – square.
- xx 63 can only be completed to another square by a number ending on 1 to a square ending on 4.
- Moreover this small square has to be 0 (mod 9) and 1 (mod 16), to obtain a square 4 (mod 9) and 0 (mod 16).
- First candidate $1^2 = 1$ (mod 9) so it drops out.
- Second candidate $9^2 = 81 = 0$ (mod 9) and 1 (mod 16).
- 7663 + 81 = 7744 which is a full square.
- We now have the "missing" prime numbers, 88 + 9 = 97 and 88 – 9 = 79.
- Final answer: 283531 = 37 × 79 × 97.

The first requirement for a successful execution of this activity is an above average knowledge of the primes, in my feeling the 3 figure primes are the minimum. If for the reader this a shot over the mark, then use the table 2 with the three figure primes up to 1,000. There are 143 three digit primes. The marked ones amongst them are the prime numbers which are the sum of two different squares in one way, one even basic number and one odd basic number. This means that all these numbers are 1 modulo 4, as an even number squared is automatically 0 modulo 4. An odd number squared results automatically in a number 1 mod 4 so that the sum of them is 1 mod 4. Examples: 149 = $10^2 + 7^2$; 373 = $18^2 + 7^2$; 881 = $25^2 + 16^2$. There are 69 1 mod 4 primes amongst the 143 other ones. The white prime numbers are all 3 mod 4.

Besides that a profound knowledge and experience with the squares is required. If this knowledge is not there, then use a table with the squares up to 1.000. Experience with the (a + b) × (a – b) formula is indispensable. And not to forget, being confident with modulo calculation.

At times one can read in internet forums a question to learn a lot of prime numbers by heart, be it up to 100, 1,000 of even 10,000. But why? There is nothing against learning by heart whatsoever, the question is: What use is made of this? If the learning is not followed by an intensive practise, the well-known saying becomes valid "If you don't use it, you lose it". And all the efforts are made in vain.

12.1 Finding two primes, bigger numbers

149137 is the product of which prime numbers, a and b? The number is composed as (a + b) × (a – b). We determine the mod 9 and mod 16 of this number, they are respectively 7 (mod 9) and 1 (mod 16). What is the shape of the small square we have to add to obtain the big square which is also 7 (mod 9) and 1(mod 16)? The small square has to be 0 (mod 9) and 0 (mod 16), so divisible by 144.

There are no other possibilities for a combination. The calculation with mod 9 and mod 16 has been chosen because of the reduction of the number of candidates.
Besides, we know that the basic number of the square to be added has to end on 2 or 8 to find a square ending on 1, as known, with an even 10. Sqrt. of 149137 = 386+.

Candidates for the small square: 12^2 + the number gives 149281, not a square. 48^2 = 2304 + the number gives 151441, no square. 72^2 = 5184 + 149137 = 154321, no square. 108^2 = 11664, + 149137 results in 160801 = 401^2. This is what we are looking for. The next step is subtracting and adding 401 – 108 = 293 and 401 + 108 = 509. So 293 and 509 are the numbers we looked for.

342557 This number is 8 (mod 9) and 13 (mod 16). So the completing square has to be 1 (mod 9) and therefore the base number is either 1 or 8 (mod 9). And the square has to be 4 (mod 16) so the basic number is only divisible by 2. The question number ends on 557, which means that the square to be found has to end on 4, with automatically an even ten, to complete to a square ending on 1 with an even ten, so over more the basic number has to end on 2 or 8. The question number is also 13 (mod 16) so we need a square 4 (mod 16) to complete to a square 1 (mod 16). Finally we have to find the big square being 0 (mod 9); the basic number ending of the small square has to end on 2 or 8 and must not be divisible by 4, so only a "bold" even number.

The square root of the question number is 585+, as 585^2 = 342225. Candidates for the big square for the time being: 591, 609, 621, 639, 651, 669, each being increased by 30 or a multiple of it.

For the small square, ending or 2 or 8 and either 1 or 8 (mod 9) our first candidate is 62, then 82, 98, 118, 262, 278. 62^2 = 3844 + qn (question number) 342557 = 346401, no square. 82^2 = 6724 + qn = 349281 = 591^2. Now we do 591 – 82 = 509 and 591 + 82 = 673, so 509 and 673 are the numbers we looked for.

Resuming, it is to be said that dependent upon which thought firstly falls in, one can look after the small square to complete or if this is easier, to look after the big square. In fact both the modulo 9 and 16 calculations are also used here.

12.2 Finding three primes of nine digit question numbers

Before starting with the real work it is good to consider some elements:

1. There are in total 143 three figure prime numbers
2. As the product of 3 of them has to result in a nine figure number, means that if one of them is small –e.g. 167 – this means that the product of the two other ones has to be at least 600,000.

After a lot of thinking and consultation with my good friend Jan van Koningsveld the following technique was developed:

	211	307		503				907
103	223	311		509	607	709	811	911
107	227		419					919
				523		727	823	
		331	431		619		827	
127	239			547	631	739		
131		347	439			743	839	947
	251		443	563	643	751		
139					647			967
	263	359		571			859	971
151		367			659		863	
	271		463	587				983
163		379	467			787		991
167		383	479	599			883	
	283		487		683		887	
179			491		691			
			499					
191								
199								
21	**16**	**16**	**17**	**14**	**16**	**14**	**15**	**14**

a. Take the next bigger prime number after the three first figures of the question number. E.g. If the question number starts with 343, then take 347 as the first prime number.

b. Take the rounded cubic root of the number. Following this method implies that anyhow one of the three prime numbers lies between these limits.

About table 2: the bold printed primes are one time the sum of two different squares and by consequence 1 mod 4, the other ones are 3 mod 4.

To be precise: not all the numbers can be used. The smallest possible number is 991 × 997 × 103 = 101,766,781. This means in practice that we will not retrieve the number 101 in our question numbers.

As our first example we take the "Dase number" 278,353,657.

The next prime number after 278 = 281. The rounded cubic root of 278353 = 65. So between 281 and 653 lies one of the three primes we are looking for. Between 281 and 653 there are 60 prime numbers, which implies that the maximum number of divisions will be 60. Then we start with "rough" division work.

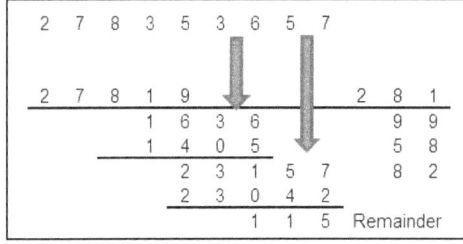

Table 3

You'll see in this way we continue until a prime number is found for which the division is an integer one, so there is no remainder. In the division by 281 we find as the first figures of the answer 99.

Subtracting 27835 − 27819 gives a remainder of 16. Then we append 36 and get 1636 ÷ 281 gives 05 times and a remainder of 231. Answer so far on the division 9905. Now we append 57 and have 23157, and divide this number by 281 which results in 82 × and a remainder of 115. This is not an integer division so we continue with the next prime number 283.

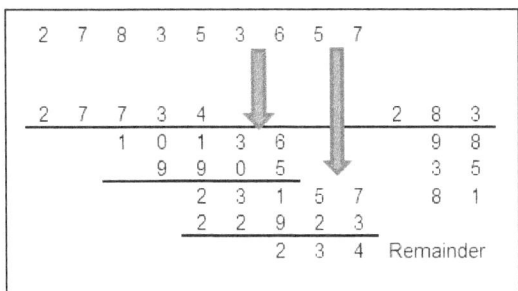

Table 4

The first division gives an answer of 98 × and a remainder of 101, and we append 36 so that the new dividend is 10136. We can subtract 35 times, i.e. 9905, and the subtraction results in a remainder of 231. Answer so far 9835.

Now we append 57 and get as a dividend 23157. Here we can subtract 81 × 283, which results in a remainder of 234. Answer so far 983581, and a remainder of 234. Neither is this an integer division, so we have to go further with the next prime number.

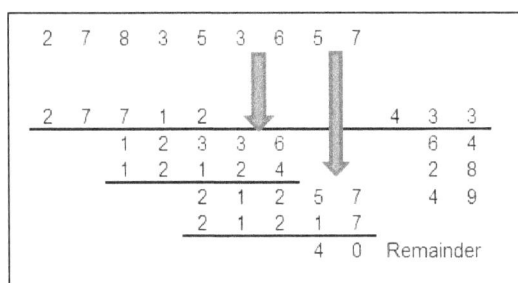

Table 5

Continuing with our divisions we have in the meantime made 24 fruitless divisions, i.e. divisions with a remainder.

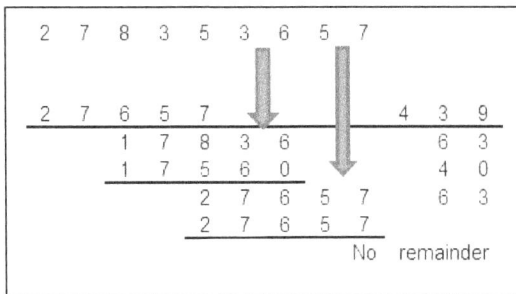

Table 6

Exactly in the same way as we did before we now have the prime number 433 which again is a division with a remainder, so what we have to do is continuing, and take the next prime number 439.

Finally, after 25 divisions, we find with the prime number 439 the result we were so intensively looking for: the complete result of the division by 439 is 634063.

So what we have to do now is to find the two other prime numbers of which 634063 is composed.

We follow here the same procedure as to be found in the chapter "Factorisation".

This number 643063 is 4 (mod 9) and 7 (mod 16), to be completed with a small square 0 (mod 9) and 9 (mod 16) to obtain the bigger square 4 (mod 9) and 0 (mod 16), of which the basic number will be divisible by 4.

The candidates are: 9, 15, 33, 39, 57, 63 each of them additionally increased by 72. The basic number of the small square has to end on either 1 or 9. This results in a square ending on 1, always with an even ten. There are no other possibilities. E.g. a square ending on 6 with always an odd ten, would result in a number ending on 9 with an odd ten. This kind of number cannot be a square. A square ending on 4 would result in a number ending on 7, which as you know, also cannot be a square.

The basic number of the big square has to end on either 2 or 8 and should be 2 or 7 mod 9. The nearest basic number 0 (mod 9) is 810 and this plus or minus 2 gives us 808 and 812. The squares are respectively $808^2 = 652864 - 634063 = 18801$, which is no square. The next candidate is $812^2 = 659344 - 634063 = 25281$, which is the square of 159. 812 − 159 = 653 and 812 + 159 = 971. Now the answer is complete: the number 278,353,657 is the product of 439 × 653 × 971.

As you can imagine, there is an intense search for an algorithm for finding the first prime number in one go, mental calculators and mathematicians are vying with each other as to who should be the first to find it, but until now it has all been in vain. One could say finding the first prime number has a high grade of gambling. Table 10 gives you an impression.

12.3 When to stop dividing

Table 7

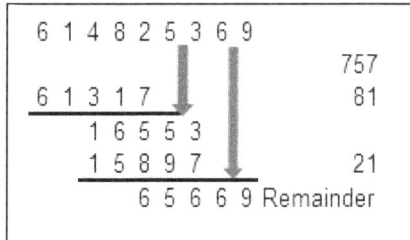

The cube root of 614,825 is 85+; nearest lower prime is 607, so one of our candidates lies between 607 and 853. After dividing by 757 81 times and then 21 times, we have a remainder 656 and append 69 to get 65669. Now we have to realise that xx69 divided by 57 is yy17. And 17 × 757 = 12869, far less than the 65669, so we do not spoil our energy in executing this division. This would consume a lot of time and bring nothing. The factorisation of 614,825,369 is 701 × 907 × 967.

Table 8

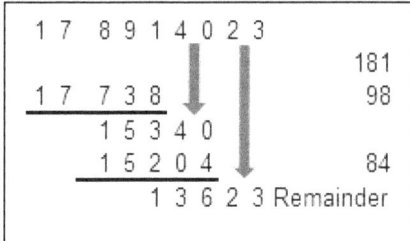

The cube root of 178,914 is 56+, nearest lower prime is 557. 179: nearest lower prime is 179, so one of our candidates lies between 179 and 557. In the same way this division. After having divided 98 times and then 84 times there is a remainder of 13623. Now we realise is 23 ÷ 81 results in 83. As 83 × 181 is far more than 13623 we do not execute this division. This would also be a waste of time.

Table 9

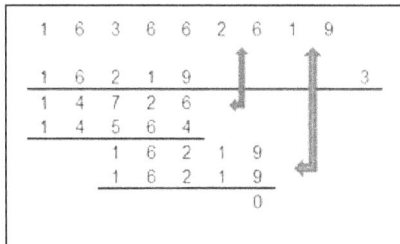

In 16366 we see 331 will go 49 times, and the result 16219 is subtracted from the question number, the remainder is 147 and 26 of the qn is appended to get 14726.

In 14726 we see 331 will go 44 times, result 14564 which is subtracted from 14726, the result is 162.

Now we append 19 to get 16219, in which go 49 times 331 with a zero remainder, so this is an integer division.

Now we are going to find out which two three digit prime numbers compose 494449.

494449 = 7 (mod 9) and 1 (mod 16) which means that we are looking for a square 0 (mod 9) and 0 (mod 16) to add to 494449 to get another square 7 (mod 9) and 1 (mod 16). The candidates herefore are 12, 24, 36, 48, 60 eventually each of them + 72. Besides the small square has to end on either 00 to get a bigger square ending on 49 or a small square ending on 76 to get a bigger square ending on 25. This means that some drop out because their square ends neither on 00 nor on 76.

A quicker approach is looking for the candidates of the big square, for which in the 700 series the candidates are: $707^2 = 499849$, $743^2 = 552049$, $715^2 = 511225$, and $725^2 = 525625$. The other squares, of which we subtract 494449, result in 5400, no square; alternatively 57600, the square of 240. Now we have 552049, the square of 743. Next step: 743 ± 240. Now we have the answer: $494449 = 503 \times 983$.

The complete answer: $163662619 = 331 \times 503 \times 983$.

12.4 Time needed

Table 10 hereunder informs us about the way of working.

Question Number	Smallest Prime	Biggest prime	Max number of Primes	Divisions	1st found	Prime 2	Prime 3	Time	Average	Average + and -
114662819	127	487	63	20	229	587	853	20	720	133
137793241	139	509	64	59	487	523	541	60	532	9
163662619	167	547	63	27	331	503	983	30	743	240
184033021	191	569	62	37	331	613	907	40	760	147
193294349	197	577	61	48	503	571	673	50	622	51
274386547	277	647	61	12	353	857	907	13	882	25
326810647	331	683	57	17	433	797	947	20	872	75
472315553	653	773	47	19	653	821	881	20	851	30
483710407	491	787	46	30	677	811	881	30	846	35
487447859	491	787	46	21	647	827	911	23	869	42
558586069	563	823	40	33	599	941	991	35	966	25
576097799	587	839	40	34	809	829	859	36	844	15
602723227	607	839	36	20	739	823	991	20	907	84
686687009	701	887	29	5	953	769	937	5	853	84
765478093	773	919	21	12	859	941	947	13	944	3
812230291	821	937	27	3	983	907	911	3	909	2
941589731	947	983	6	4	953	991	997	5	994	3

Please: before starting to calculate, have a look at this table. It will undoubtedly strike you that in the lower numbers there is an awful number of candidates, to be found in the column "number of primes".

Let's take the first number 114662819. 127 is the smallest possible prime number after 114 and 487 is the rounded cube of the whole number. Between these numbers there are no fewer than 63 prime numbers and until now there is no algorithm which enables us to find the first prime more quickly than to start with "rough" dividing.

Inevitably the question is raised if there is any algorithm for finding in one go the first prime number. My answer is NO! For if this were possible this would mean the end of the elusiveness of the prime numbers and the essence of their existence. And their role in e.g. the encryption of electronic messages and whatever more would be finished. If ever there should be found an algorithm for finding three or even two figure prime numbers, a step to the bigger numbers would be a matter of – probably even short – time.

And then we see that, after dividing twenty two times, 229 gives us the first prime number, as 114662819 ÷ 229 results in an integer answer; for all clarity it is a division without a remainder; the result is 500711. As the number of divisions is given as 20, we can conclude that the beginning of the search was with 127, after that 131, 139, 149 et cetera.

Dividing by 229 results in 500711 = 5 (mod 9) and 7 (mod 16). This number should be completed to a "big" square which has to be 0 (mod 9) and 0 (mod 16) by a smaller square which is 4 (mod 9) and 9 (mod 16). The number 500711 could be completed by a smaller square ending on 25, of which the hundreds are without any exception always even to a big square ending on 36 with an odd hundred. The other possibility is completing with a smaller square ending on 89, of which the hundreds without exception are always even, to the big one ending on 00 with an even hundred. Sqrt 500711 = 707+.

We'll try the 36 possibilities. The first candidates after 707 are 744^2 = 553536 and 756^2 = 571536. This is a blind alley, as this requires the hundreds of the xx25 squares to be 8, which does not exist. The first 00 candidate is 720^2 = 518400, and this is a full hit: 518400 − 500711 = 17689 = 133^2. We can now complete our efforts easily: the numbers required are 720 − 133 = 587 and 720 + 133 = 853. In the column "average" we find 720 and in the "+ and −"column we find 133.

Final answer: 114662819 = 229 × 587 × 853

For the − big − difference we'll have a look on the last number of the table: 941589731. Smallest prime 947, biggest prime 983, "distance" between them six prime numbers, which means no more than six divisions. The table shows us that only four divisions do. 941589731 ÷ 953, the first prime number with a zero remainder = 988027 = 7 (mod 9) and 11 (mod 16). This number has to be increase by a smaller square 0 (mod 9) and 9 (mod 16) to obtain a bigger square 7 (mod 9) and 4 (mod 16). 988027 + 3^2 = 988036 = 994^2. Now we have 991 and 997, the complete answer of the factorisation of 941589731 = 953 × 991 × 997.

Having a look at the column "divisions" we see the smallest number is 3 divisions with 27 prime numbers, question number 812230291: the biggest number is 59 divisions with 64 prime numbers with the number 137793241. This depends on the way of working: starting with the biggest possible prime number or the smallest one. So if I had begun with 509 and then counted down, already after three divisions there was the result. This is the gamble. If we take an average of the time needed we find a rounded time of 25 minutes.

12.5 Where to start??

Now raises the question: where to start in the case of a big number of divisions.

We take again the number 114662819, where the lowest candidate is 127 and the highest is 487.

One can:

- Start with 127 and continue to 487, maximum 63 divisions
- Start with 487 and go back to 127 until the moment of a "full hit", a division without a remainder, again maximum 63 divisions
- Start with 127 and end with 199, the last prime number in the 100 series
- Then start with 487 and work backwards to 401, the lowest prime number I the 400 series
- Then work from 211, the first prime number in the 200 series and end with 293
- Start with 397 and work backwards to 307, the lowest prime number in the 300 series.

After all, which is always very easy, one can say that the first option – starting with 127 and still going strong – would result in the fastest integer division, after twenty divisions.

Again, a ready knowledge of the primes up to 1,000 is required to solve the questions. Regardless what one does, there is no superior way of working; there is only the "Russian Roulette".

About the prime numbers up to 1,000: We shall use the table for them.

Immediately it will strike you that there are a lot of prime numbers each in a gray box. They are all 1 mod 4, so the sum of two different squares in one way; an even one and an odd one. E.g. $157 = 11^2 + 6^2$; $457 = 21^2 + 4^2$; $853 = 23^2 + 18^2$.
The prime numbers in white boxes are all 3 mod 4, so they cannot be the sum of two different squares.

There are 143 three figure prime numbers, of which 69 are the sum of two different squares in one way; roughly said, about half.

The other prime numbers are 3 mod 4.

If we renounce the 21 prime numbers in the 100 series, again, roughly said the average number of primes is 15 per hundred.

In the case of a higher number it can occur that another approach, starting with 997 and then working backwards, results in less work; there are fewer divisions to be made before there is an integer answer.

The number 686687009 has a lowest candidate 701 and highest 887, 29 prime numbers "distance". Anyhow, one of the prime numbers between 701 and 887 results in an integer division. But, and we should never forget that: this means that at least one of the other prime numbers we are so desperately looking for, lies between 997 and 887. And from 997 working backward to 887 there are no more than 14 prime numbers, so this is the shorter way to our goal.

Working this way we found that 686687009 ÷ 953 results in 720553 as an integer division, so this number, 4 (mod 9) and 9 (mod 16) has to be completed to another square, in this case this is 84^2, to obtain 727609 = 853^2. Final answer then is 953 × 769 × 937, found after only five divisions.

12.6 Time needed?

My general average is 1 minute for a division, disregarded if the answer is an integer number or a remainder. To avoid unnecessary efforts it is a matter of seeing quickly if a division is integer or leaves a remainder.

12.7 How many numbers of this kind are there?

Of course the question arises of how many combinations can be made with three figure primes under the conditions; they are not consecutive and there are no squares, i.e. in one question a number cannot appear twice.

The answer comes from my good friend Dr. Andy Robertshaw; he is a promoted mathematician.

For information, the number of combinations of 3-digit primes without repeats is 143 x 142 x 141 / (3 x 2 x 1) = 477191. However this allows for consecutive prime numbers to be included.

Now let us not allow consecutive primes.
There are 140 three-digit primes not being used. If we have 140 blocks in a line to represent the unused primes, we can insert three blocks into spaces between the 140 already in place. We can also allow one of our three blocks to go at each end of the 140. So there are 141 possible places we could insert each block.

Once the 3 new blocks are in place, we can then allocate a number to each block from left to right. We use the three-digit prime numbers in ascending order. The three new blocks will represent the three prime factors.

Since there are 141 places for each new block, there are 141 x 140 x 139 / (3 x 2 x 1) possible arrangements = 457310. This is the number of possibilities without repeats or consecutive primes in the problem.

12.8 How to get the question numbers

A basic rule in the mental calculation world is "Never make your own questions", because otherwise you have already the lion share of the answer when starting calculation. Mr. Steve Wolfram was so kind to help me with a license for his formidable calculation and academic mathematic program "Mathematica" which generates question numbers in a flash.

The instruction for Mathematica contents in general are the following instructions:

- The number of question numbers wanted
- Random Prime 100,999
- $A \neq b \wedge a \neq c \wedge b \neq c, x = a\,b\,c$
- If $x > 100000000$, Append To List

Which means that the number of question numbers should be given, the prime numbers are taken at random, they are all different and that the multiplication results in a 9 digit number.

12.9 Analysis

The "philosophic" aspect

After having factorised so many numbers the question came up: is there not an algorithm to be found with which the first prime numbers quickly can be found? Then two choices struggled for the priority: the ambition to find one and the down-to earth logic.

The last one says: if ever such an algorithm should be found, this would mean the end of the mystery of the prime numbers and it would no longer be possible to use prime numbers for the encryption of electronic messages. And our safety would no longer be guaranteed. For if it were possible to find a three digit number quickly, the step to find bigger prime numbers would only be a small step.

The facts

We take as an example the table, first number 114662819, last number 941589731, in total 100 "Dase" numbers. We now make a subdivision in the "hundreds" series:

From	Numbers	From	Numbers
101.....193	21	604.....686	7
202.....292	13	702.....789	8
306.....397	13	808.....888	8
402.....498	12	913.....	6
501.....577	9		

Table 11

What is the logic behind all this? Amongst other things, this: The decisive condition is that the question number is the product of three 3 digit prime numbers. So for a number in the 300 series all the prime numbers smaller than 300 automatically drop out, so quite simply the number of possibilities reduces.

This also explains that in the lower numbers the "distance" between the lowest and the highest prime numbers can even be 67 prime numbers: for the question number 128098081 the "distance" between the lowest prime number 131 and the highest prime number 503 is 67 prime numbers!!

12.10 The "Russian Roulette"

There is no reasonable advice on where to start, because the first prime number we need in question number 128098081 can be anywhere between 131 and 503. If one starts with 131 and counts up, then after 35 divisions there is one without a remainder, by 337, answer 380113. (The finding of the two other factors has been described earlier in this chapter). If we started with 503 and then counted down, after 28 divisions we arrived at 337. One could also start with 137, count up to 199, then start with 503 and count down to 401 and continue with 211 up to 293, then count down from 397 to 337, in total 54 divisions. Again: however challenging this kind of question may be they have a high "Russian Roulette" percentage.

In fact this kind of question is not suitable for a world record attempt; the time needed for a solution varies from 1 minute up to 63 minutes, and after 100 of these numbers there is an average time of 19.9 minutes. For only one number it is a matter of luck, for so many numbers that there is a reasonable average, but there is not enough time: even if one takes only 20 of these "Dase" numbers, with an average of 19.9 minutes, this requires a non-stop effort of over 6 hours of intensive calculation. Who has that much stamina? Let alone that somebody has so much energy that he / she can solve so many questions in one go. As I succeeded in factorising the number 210,681,151 in 397 × 601 × 883, I was awarded the "Prix d'Excellence" in October 2014.

In fact this activity is a kind of factorisation, although generally the numbers offered in the competitions to be factorised are composed of a lot of prime numbers, and their values do not exceed 100.

13 Modulo calculation, introduction

Modulo calculation is a form of calculation with integer numbers with a number which acts as the upper limit, the modulus.

In modulo calculation we work with a number M, we start with 0, 1, 2, ….m–1. After that we do not get m, but start again with 0. The result of a modulo calculation is the remainder of a common calculation after dividing by the modulus m. The standard definition of any operation is used, but if the result is bigger than M, so many times M is subtracted until the result is smaller than M. Numbers which are mod M equal, so which differ by a multiple of M, are called congruent modulo M.

In this chapter you will see that modulo calculation is a very good expedient in a wide variety of arithmetic operations.
We take as an example mod 9. A number divided by 9 with a remainder 4 can also be written as 4 (mod 9).
The proof for the correctness of modulo calculation – this does not come from the author – is the following.
Let us take two numbers, which we want to multiply mod 9: p + 9s and q + 9t. Multiplying (p + 9s) × (q + 9t) gives: pq + 9(pt + qs) + 81st. For divisibility by 9 we can take the nines out. What remains is pq. So (p + 9s) × (q + 9t) gives the same 9 remainder as p × q. By doing like this we spare ourselves the effort of multiplying big numbers.
You'll understand that with another modulus it works in the same way.

Modulo calculation can help us in practically every kind of calculation operation for finding the correct answer or checking if the answer we found is indeed correct. It is faster than to repeat the operation which we did and possibly make the same error again.

The modulo of any number is always lower than the modulus itself:
In fact in modulo calculation we only work with remainders, the coefficient is of no importance. So 15, 24, 33 are all 6 (mod 9), and according to the definition congruent mod 9.

It is practical to take a modulo which is easy to use. However: the bigger the modulus, the bigger the "distance" between two congruent moduli: 6, 15 and 24 are all 6 mod 9. If we take modulus 101 within the same hundred there is only one 6 (mod 101).

As examples of moduli which are more or less easy to work with:

- 7, in fact calendar calculation is modulo 7 calculation
- 9 because the mod 9 is so easily to be found: divide the sum of the individual figures of the number, take the remainder and you have the mod 9.
- 11, subdivide a number from right to left in groups of two figures, add the individual 11 remainders, divide by 11 and there is the mod 11
- 13 because $7 \times 11 \times 13 = 1001$, just $1000 + 1$
- 16, more about 16 in the chapter Factorisation.
- 27 as $27 \times 37 = 999 = 1{,}000 - 1$
- 37 as $37 \times 27 = 999 = 1{,}000 - 1$
- 41 as $2439 \times 41 = 99{,}999 = 100{,}000 - 1$
- 73 as $73 \times 137 = 10{,}001 = 10{,}000 + 1$
- 91 as $11 \times 91 = 1001$, $10989 \times 91 = 999{,}999$
- 101 as it is $100 - 1$ or $10{,}000 = 99 \times 101 + 1$
- 137 as $137 \times 73 = 10{,}001 = 10{,}000 + 1$

These numbers will be discussed separately.

13.1 Restrictions

Is every modulus at all times or in all kinds of operations to be used? No. During working with these numbers I was confronted with some unwelcome surprises. Modulo calculation can be very helpful, depending on which modulus is taken for which kind of operation.

13.2 The moduli

7

The number 7 is often combined with 11 and 13, as $7 \times 11 \times 13 = 1001$. In the case of factorising a six digit number with "symmetry" e.g. 162162, one can immediately write 7, 11 and 13 as the first figures of the answer. Remaining is 162 which is simple to factorise as 2 and 3^4, so the complete answer $162162 = 7 \times 11 \times 13 \times 2 \times 3^4$.

Or: 6,522,516. You'll quickly see $6522 - 516 = 6006$, therefore the number is divisible by 1001, so now we have already 7 × 11 × 13. Then factorise 6516, which is $2^2 \times 3^2 \times 181$. The complete answer is 6522516 = 7 × 11 × 13 × $2^2 \times 3^2 \times 181$.

13.2.1 Calendar calculation, introduction

Furthermore the number 7 is indispensable in calendar calculation, where some numbers have to be added and the modulo 7 remainder represents the weekday. In fact calendar calculation is nothing else than modulo 7 calculation.

For the calculation of a weekday we have successively to deal with:

The date
The month
The year
The century

Warning: there are different ways of counting and calculations of the offsets, so do not mix up the various methods!!!!

What happens is this: There is an addition of the elements; the date, perhaps 13, the month, let's say February, the year, e.g. 79, and the century e.g. 1900.

We'll go through it in detail.

The weekday

1	Sunday	5	Thursday
2	Monday	6	Friday
3	Tuesday	7 or 0	Saturday
4	Wednesday		

As any week contains seven days this will not give us problems: one can randomly add or subtract a multiple of 7 and the weekday remains the same. If the 18th of any month is a Tuesday, the 4th, the 11th, and the 25th are too.

The month

For the month there is an offset:

January	1	April	0	July	0	October	1
February	4	May	2	August	3	November	4
March	4	June	5	September	6	December	6

The way these offsets are calculated is explained in the chapter "Calendar calculation". To be complete: In leap years, for the month January and February 1 day has to be subtracted – do not forget that!!

The year

For the years there are two calculations required, they are to be divided by:

4, to calculate the number of leap years
7, to calculate the remainder after division by 7

E.g. year 13. Herein fit three times four, and when divided by seven the remainder is six. Total offset for the year 13 is $3 + 6 = 9 = 2$ (mod 7).

Year 45: herein fit eleven times four and when divided by seven the remainder is 3, total offset of the year $45 = 11 + 3 = 14 = 0$ (mod 7).

The century

The years of a century are counted as leap years if the number is divisible by 400. This was the case in the years 1600 and 2000. But for example the years 1700, 1800 and 1900 were not leap years.
For the 20th century, the years 1900, there is no addition.

One century after the 20th we add 6 as 100 years include 25 leap years and 100 divided by 7 gives a remainder 2, $25 + 2 = 27 = 6$ (mod 7). Therefore for the 21th century 6 is added.

Further on in this book a special chapter is devoted to calendar calculation.

9

A practical modulus to work with. We only need to add the individual figures of a number, divide the result by 9 and then have the mod 9.

The great attraction of the number 9 is:

- The mod 9 of a number is easily to find, one takes the sum of the individual figures and divides the sum by 9 and here you are! E.g. $743825 = 7 + 4 + 3 + 8 + 2 + 5 = 29 \div 9 = 3$ r 2. The quotient is – according to the definition – not mentioned. We write shortly $743825 = 2$ (mod 9).
- One always works with small numbers.

Nine is a very practical number to work with. For determining the mod 9 of a given number it will do to add the individual figures of that number and divide the result by 9, and here you are!

13.2.2 An addition

We start with a simple addition: 48 + 65 + 91 + 62 + 43 + 59 = 368. We can check the correctness of the answer by modulo calculation.

When we do the 9-test we determine the 9-moduli of the individual numbers, add them, determine the 9-modulo of this addition and compare with the answer we had already.

So 3 + 2 + 1 + 8 + 7 + 5 = 26, = 8 (mod 9). As 368 = 8 (mod 9) too, we may assume that 368 is the correct answer.

Nevertheless, some caution is required. If the nines test does not agree with the answer, it is for 100% sure that the answer is incorrect. Still there is a little chance that the answer is wrong. We take: 346 + 718 = 1163. This answer is wrong, but the nines test gives a good result: 3 + 4 + 6 = 13 = 4 (mod 9); 7 + 1 + 8 = 16 = 7 (mod 9); 1 + 1 + 6 + 3 = 11 = 2 (mod 9)

This is pure theory; from my calculation practice I do not remember a case like that.

13.2.3 A multiplication

```
        67         6 + 7 = 13 = 4 (mod 9)
        89 ×       8 + 9 = 17 = 8 (mod 9)
       ────
       5963        5 + 9 + 6 + 3 = 23 = 5 (mod 9)
```

Test 4 × 8 = 32 = 5 (mod 9)

As the remainders are in accordance you may suppose your result to be correct. The mathematician will say "There is a chance of 1 out of 9 that nevertheless the result is not correct". This means that the bigger the modulus, the less chance there is of an error. But also the bigger the modulus, the more it is difficult to work with.

Should you be more confident with the 9-test you'll see with 5963 that you can ignore the 9, and as 6 + 3 = 0 (mod 9) too, so you only have to consider the 5.

Now the same 9-test with bigger numbers.

```
       6541        6 + 5 + 4 + 1 = 16 = 7 (mod 9)
       2879 ×     2 + 8 + 7 + 9 = 26 = 8 (mod 9)
      ────────
      18,83,15,39  Mod 9: 1 + 8 + 8 + 3 + 1 + 5 + 3 + 9 = 20 = 2 (mod 9)
```

When you have more experience you'll immediately see the 18, which can be ignored, and the 15 + 39 = 54 = 0 (mod 9), so you only have to consider the 83, which is 2 (mod 9).

After getting more confident with MC, you'll see in 6541 that you can ignore 5 + 4, being 0 (mod 9), so the number is 6 + 1 = 7 (mod 9).

In 2879 you can ignore the 9 himself and 2 + 7 = 0 (mod 9), only the 8 remains, so 2879 = 8 (mod 9).

The test becomes 7 × 8 = 56 = 2 (mod 9)

Do some multiplications and check the results with the 9-test. Of course you can firstly do the nines test and then check your answer, whichever you prefer.

Questions

To calculate and check with the nines-test:

$$68 \times 97 \;;\; 374 \times 821 \;;\; 1637 \times 8251$$

My teacher from many, many years ago told us there were a risk with the 9-test. If in the result of the multiplication two digits would be interchanged, the 9-test would not reveal that. A surprising idea led me to the 11-test, which I bethought myself. It is simple to see: 24 and 42 are each 6 (mod 9), 24 = 2 (mod 11) and 42 = 9 (mod 11). Many years later I experienced that the elevens test existed already a long time. It is not clear how probable it is that the interchange will happen. In the daily practice the chance that interchange of two digits will happen is negligible, and the chance that it will not be discovered is still smaller.

Answers

68 × 97 = 6596. Nines test: 68 = 5 (mod 9) en 97 = 7 (mod 9). 5 × 7 = 35 = 8 (mod 9). 6596: 6 + 5 + 9 + 6 = 26 = 2 + 6 = 8 (mod 9). The nines remainders agree.

374 × 821 = 307054. Nines test: 3 + 7 + 4 = 14 = 1 + 4 = 5 (mod 9); 821 = 8 + 2 + 1 = 11 = 1 + 1 = 2 (mod 9). 2 × 5 = 10 = 1 + 0 = 1 (mod 9). 307054 = 3 + 0 + 7 + 0 + 5 + 4 = 19 = 1 + 9 = 1 (mod 9). The nines remainders agree.

1637 × 8251 = 13506887. Nines test: 1 + 6 + 3 + 7 = 17 = 1 + 7 = 8 (mod 9). 8 + 2 + 5 + 1 = 16 = 1 + 6 = 7 (mod 9). 8 × 7 = 56 = 5 + 6 = 11 = 1 + 1 = 2 (mod 9). 13506887 = 1 + 3 + 5 + 0 + 6 + 8 + 8 + 7 = 38 = 3 + 8 = 2 (mod 9). The nines remainders agree.

13.2.4 A division

In a division it is a bit more intricate. What are we to do if we have to calculate a division with the dividend 2 (mod 9) and the divisor is 7 (mod 9)? We do this: We take 2 and add as many times as needed 9 (the modulus) until we have a number which is divisible by 7. Then we get: 11, 20, 29, 38, 47, 56. Here we can stop because 56 is divisible by 7, namely 8 ×. So a division with the dividend 2 (mod 9) divided by a number which is 7 (mod 9) results in an answer which is 8 (mod 9).

$$35696 \div 97 = 2 \text{ (mod 9)} \div 7 \text{ (mod 9)} = 8 \text{ (mod 9)}.$$

As 291 < 351 < 388 = First digit = 3. 6 ÷ 7 = 8 uniquely. Answer so far 3 ? 8. As 3 + 6 = 0 we need to add 6 to get 8 (mod 9), so final answer 368.

Questions

$$7178 \div 97 \,;\, 28397 \div 73 \,;\, 112751 \div 137.$$

Warning: in the case of a division where both the dividend and the divisor have the same modulus, disregarded which one you take, you have to change to another modulus.

E.g.: 1728 ÷ 18 → in cases like this you cannot work with modulo 9, as any number multiplied with 9 results in an answer 0 (mod 9). In this question you can take refuge in modulo 11.

17 28 = 1 (mod 11); 18 = 7 (mod 11). 1 (mod 11) ÷ 7 results in 8 (mod 11). And indeed, the answer 96 = 8 (mod 11).

Answers

7178 = 5 (mod 9). 97 = 7 (mod 9). 5 + 9 = 14 ÷ 7 = 2 (mod 9). The answer has to be 2 (mod 9). Next: ?8 ÷ ?7 gets 28 ÷ 7 = 4. The last figure is a 4. Answer so far ?4 = 2 (mod 9). There has to be added 7 to arrive at 74 = 7 + 4 = 2 (mod 9). Answer: 7178 ÷ 97 = 74.

28397 = 2 + 8 + 3 + 9 + 7 = 29 = 2 + 9 = 2 (mod 9). 73 = 7 + 3 = 1 (mod 9). 2 ÷ 1 = 2, the answer is 2 (mod 9). ?7 ÷ ?3 gets 27 ÷ 3 = 9. So the last figure of our answer is 9. 3 × 73 = 219 and 4 × 73 = 292, so the first figure of our answer is 3. We now have 3 ? 9 = 2 (mod 9). This fits only if in the place of the question mark there is an 8: 389 = 3 + 8 + 9 = 20 = 2 + 0 = 2 (mod 9), the answer is 389.

112751 ÷ 137. 1 + 1 + 2 + 7 + 5 + 1 = 17 = 1 + 7 = 8 (mod 9). 137 = 1 + 3 + 7 = 11 = 1 + 1 = 2 (mod 9). 8 ÷ 2 = 4, so the answer is 4 (mod 9). ?1 ÷ ?7 gets 21 ÷ 7 = 3, the last figure of our answer. In 112 we can fit 8 × 13 + something = 104 + something. We now have 8 ? 3 = 4 (mod 9). 8 + 2 = 11 = 1 + 1 = 2 (mod 9). To arrive at 4 (mod 9) there has to be added 2, and now we have 823 = 8 + 2 + 3 = 13 = 1 + 3 = 4 (mod 9), and so 112751 ÷ 137 = 823.

13.2.5 An integer square root

$\sqrt{5776}$: 7 (mod 9). The answer has two digits. First digit of the answer 7, as 49 < 57 < 64. The second one is either 4 or 6. 74 = 2 (mod 9), 2^2 = 4 (mod 9). 76 = 4 (mod 9) 4^2 = 7 (mod 9) so 76 is the correct answer.

Mod 9 exponents 1, 2, 3 and 4

	1	2	3	4	5	6	7	8
1		2	3	4	5	6	7	8
2	2		6	8	1	3	5	7
3	3	6		3	6	0	3	6
4	4	8	3		2	6	1	5
5	5	1	6	2		3	8	4
6	6	3	0	6	3		6	3
7	7	5	3	1	8	6		2
8	8	7	6	5	4	3	2	

BN	BN³ (mod 9)
1	1
2	8
3	0
4	1
5	8
6	0
7	1
8	8

BN	BN⁴ (mod 9)
1	1
2	7
3	0
4	4
5	4
6	0
7	7
8	1

$\sqrt{52441}$: 7 (mod 9), so the basic number (BN) is either 4 or 5 (mod 9). First two digits of the answer 22. Possibilities 221, 5 (mod 9) and 229, 4 (mod 9). Here we are at the limits of the 9-test and have to consider that 229² is closer to 52900, 230², 229 is the correct answer. Or you can do this: 524(QN) – 484(22²) = 40. 529(23²) – 484(22²) = 45. 40 ÷ 45 = ± 0.9, therefore the answer is 229.

13.2.6 About the tables

The shaded table gives the mod 9 in multiplication, and in reverse for division. If a number 7 (mod 9) is divided by a number 2 (mod 9) the answer will be 8 (mod 9). Where the same numbers cross is the square.

For working in the third power 9 is far from practical, and for the fourth power in fact the same can be said. Conclusion: The nines test is only useful in operations of the first or second degree.

13.2.7 Missing digits question

$$1526982688 \div 389?? = 39248$$

As the dividend has 10 digits and the first digits of the divisor are higher than those of the dividend, the answer will have five digits, which is already given. Firstly think about the last two digits; 88 ÷ 48 in the 2 figure area. This means that the possibilities are 06, 31, 56 and 81.

We now determine the mod 9 of the question number: 1 + 5 + 2 + 6 + 9 + 8 + 2 + 6 + 8 + 8 = 55 = 1 (mod 9). The mod 9 of the answer number is 3 + 9 + 2 + 4 + 8 = 26 = 8 (mod 9).

1 (mod 9) ÷ 8 = 8 (mod 9), which means that the divisor has to be 8 (mod 9). 389 = 2 (mod 9) and has to be completed to 8 (mod 9). This means that on the place of the question marks has to be 06. The final answer is 38906.

11

11 Is a "bold" prime number, it is only prime and not the sum of two different squares. This makes the number very attractive. In this book you'll find a lot of examples in which the calculation is made using modulo 11.

Because of its property that $9 \times 11 = 99 = 100 - 1$ it is a very useful expedient to work with. It is the only "bold" prime number with such attractive properties. Whatever I did to find a bigger number with the same properties, I did not find one so far.
For me the most practical procedure with modulo 11 calculation is to split up the given number from right to left in groups of two figures, then taking the individual moduli 11 of these groups, adding them and then determine the modulo 11 of the sum which is equal to the modulo 11 of the whole number.

E.g. 3798263 → 3,79,82,63, in mod 11 from right to left, = 8 + 5 + 2 + 3 = 18 = 7 (mod 11). In order not to fall into making mistakes, with + for the even powers of 10 and − for the odd powers of 10, from the + and − method is renounced on purpose.

Moreover, one realises that for the two figure numbers it is very simple to determine the mod 11. E.g. 25 = 22 + 3 so 3 mod 11; 74 = 66 + 8, so 8 mod 11.

Mod 11 exponents 1, 2, 3, 4 and 5

	1	2	3	4	5	6	7	8	9	10
1	1	2	3	4	5	6	7	8	9	10
2	2	4	6	8	10	1	3	5	7	9
3	3	6	9	1	4	7	10	2	5	8
4	4	8	1	5	9	2	6	10	3	7
5	5	10	4	9	3	8	2	7	1	6
6	6	1	7	2	8	3	9	4	10	5
7	7	3	10	6	2	9	5	1	8	4
8	8	5	2	10	7	4	1	9	6	3
9	9	7	5	3	1	10	8	6	4	2
10	10	9	8	7	6	5	4	3	2	1

BN	BN^3 mod 11	BN	BN^4 mod 11	BN	BN^5 mod 11
1	1	1	1	1	1
2	8	2	5	2	10
3	5	3	4	3	1
4	9	4	3	4	1
5	4	5	9	5	1
6	7	6	9	6	10
7	2	7	3	7	10
8	6	8	4	8	10
9	3	9	5	9	1
10	10	10	1	10	10

The table gives an overview of the 11 moduli in the various powers. There is a "mirror effect" in the second and fourth powers, where there are 5 pairs of numbers with the same modulus, in the 5th power we have beside 0 only 2 different possibilities, so in this case 11 is of no use. In the 1st and 3rd power there are no mirror effects.

In my calculation practice it did not happen that a given answer appeared to be wrong with the 11-test despite being the correct result.

About the tables: Where the same digits cross in the dark boxes is the square of them.
The number 11 is very useful in the first, second and third power, for the fourth power there are the same moduli as in the second power, only the order is different. For the fifth power there are only two different moduli, 1 and 10, which make the use of it a nonsense.

13.2.8 Surprise, surprise!!

Please have a look at the table BN^4 (mod 11). We also see there the mirror effect, so we have five pairs of numbers calculated to the fourth power with the same mod 11 remainder. With the other moduli such as 37, 41, 61 et cetera we see in the fourth power sets of four numbers with the same remainder. E.g. 2, 12, 25 and 35 calculated to the fourth power, so 4 numbers, have all the same 37 remainder; 16. After a lot of thinking I found out why this. 37, 41, 61 et cetera are the sum of two different squares. $2^2 + 12^2 = 0$ (mod 37), the squares are their complements, so squaring the squares we get the "mirror" effect: 4^2 and 33^2 both are 16 (mod 37).

13.2.9 A multiplication

$$
\begin{array}{ll}
6541 & 65 + 41 = 106 = 7 \text{ (mod 11)} \\
2879 \times & 28 + 79 = 107 = 8 \text{ (mod 11)} \\
\hline
18,83,15,39 & 18 + 83 + 15 + 39 = 1 + 55 = 1 \text{ (mod 11)} \\
& 7 \times 8 = 56 = 1 \text{ (mod 11)}
\end{array}
$$

It is always practical to have a look over the whole number. E.g. here you'll see 18 and 15, together 0 (mod 11) so you have only to deal with 83 + 39, 6 + 6 = 1 (mod 11).

Questions

Multiply the following numbers and check the results with modulo 11.

$$84 \times 91; \quad 397 \times 643; \quad 4653 \times 9101$$

Answers

84 × 91. 84 = 7 (mod 11); 91 = 3 (mod 11). 7 × 3 = 21 = 10 (mod 11). 84 × 91 = 7644 = 44 (0 mod 11) + 76 = 10 (mod 11). The eleven-remainders agree.

397 × 643. 397 = 3 + 97 = 100 = 1 (mod 11). 643 = 43 + 6 = 49 = 5 (mod 11). 5 × 1 = 5 (mod 11). 397 × 643 = 255271 = 71 + 52 + 25 = 71 + 77 (0 mod 11) = 71 = 5 (mod 11). The eleven-remainders agree.

4653 × 9101. 53 + 46 = 99 = 0 (mod 11). Therefore the answer will be also 0 (mod 11). 4653 × 9101 = 42346953 = 53 + 69 + 34 + 42 = 198 = 0 (mod 11). The eleven-remainders agree.

13.2.10 An integer division

$$3140146197 \div 45963$$

From right to left in groups of 2 we count: 9 + 6 + 3 + 7 + 9 = 34 = 1 (mod 11). 4 59 63 = 8 + 4 + 4 = 5 (mod 11). In the sense of modulo 11 calculation this is 1 ÷ 5 (mod 11). For finding the answer we take 1 (mod 11) and add (as many times as we need) 11 until we have a number 1 (mod 11) which is divisible by 5. We get 1, 12, 23, 34, 45 and here we are. Now we divide 45 by 5 and get 9, so the answer of the division has to be 9 (mod 11).

The first digit of the answer can be found as follows: 45963 is very close to 46(000). 6 × 46 = 276; 7 × 46 = 322, so the first digit is 6.

The last digit is 9 by uniqueness. 9 × 63 = 567, the ten is 6. For every ten more than 09, 3 has to be added. Following from 67 we get, when multiplying 63 with 19, 29, 39 et cetera, 67, 97, 27, 57. So the last two digits 19, are as 19 × 63 = ..97. The answer so far is 6xx19 = 3 (mod 11). So we miss 6 (mod 11). 31401 ÷ 460 = 68 +. We now have 6(80) 80 = 3 (mod 11), 83 = 6 (mod 11), so the complete answer is 6 83 19. Adding 6 + 83 + 19 = 108 = 9 (mod 11) so the answer stands the elevens test.

$$3203328960 \div 40864$$

The dividend: 32 03 32 89 60. From right to left in groups of 2 digits we count: 5 + 1 + 10 + 3 + 10 = 7 (mod 11). The divisor 4 + 08 + 64 = 10 (mod 11) → 7 ÷ 10 = 4, so the answer has to be 4 (mod 11).

32033 ÷ 408 = 78+. For dividing 8960 by ++864 we have these possibilities: 015, 140, 265, 390, 515, 640, 765, 890.

78000 = 10 (mod 11), and 390 = 5 (mod 11), so 78390 = 4 (mod 11) and is the correct answer.

Questions to be done with modulo 11

$$8633 \div 97 \; ; \; 124307 \div 197 \; ; \; 35336864 \div 4276.$$

Answers

$8633 = 86 + 0$ (mod 11) $= 9$ (mod 11). $97 = 9$ (mod 11). $9 \div 9 = 1$, so the answer is 1 (mod 11). ?3 ÷ ?7 = has uniquely the answer 9. Answer so far ?9 = 1 (mod 11). The only possibility: $89 = 1$ (mod 11).

$124307 \div 197 = 07 + 43 + 12 = 62 = 7$ (mod 11). $197 = 97 + 1 = 98 = 10$ (mod 11). $7 + 33 = 40 \div 10 = 4$ (mod 11), so the answer will be 4 (mod 11). Next: 124 ÷ almost 20 = 6 + something. Answer so far 6??. $7 \div 7 = 1$ uniquely, so the last figure of the answer is 1. We now have $6?1 = 4$ (mod 11). $6 + 01 = 7$ (mod 11). We miss 8 (mod 11) as $8 + 7 = 15 = 4$ (mod 11). As $30 = 8$ (mod 11) 631 is the full answer.

$35336864 \div 4276$. $64 + 68 + 33 + 35 = 200 = 2$ (mod 11). $33 = 0$ (mod 11) and $68 + 64 = 132 = 0$ (mod 11). If you are more experienced you see these elements and can take only $35 = 2$ (mod 11). $4276 = 42 + 76 = 118 = 8$ (mod 11). The mod 11 division is $2 \div 8 = 24 \div 8 = 3$ (mod 11). The answer has four figures, of which the first one will be $353 \div 42 = 8+$. The last figure: Attention please; 6864 as the last part of the dividend is divisible by $16 = 2^4$, the divisor contains 2^2, therefore the answer will end on 4 with an even ten. The answer so far $8??4 = 3$ (mod 11). Then ?64 ÷ 76 gives four possibilities: 14, 39, 64 or 89. The only one which has at least $2^2 = 64$, therefore our answer so far is now $8?64 = 3$ (mod 11). $80 + 64 = 144 = 1$ (mod 11) and if we add 2 we have $8264 = 82 + 64 = 146 = 46 + 1 = 146 = 3$ (mod 11). The final answer 8264.

13.2.11 An integer cubic root

$$2 \; 64 \; 85 \; 13 \; 43 \; 25 \; 47 \; 86 \; 75 \; 44 \; 93$$

We have twenty one figures, so the answer has seven digits, for mod 11 calculation we may and we ignore 264 and 44, as they are 0 (mod 11).

Then from right to left the 11 moduli of the two digit numbers are $5 + 9 + 9 + 3 + 3 + 10 + 2 + 8 = 5$ (mod 11), so the basic number $= 3$ (mod 11), according to the mod 11 table.

The number has 21 figures. If we take the first 6 figures, 264851, we find that the first two figures of the answer are 64, as $64^3 = 262144$; $65^3 = 274625$. Next the Newton method: $264851 - 262144 = 2707 \div 3 \times 64^2(12288) = 0.21+$. The answer so far is 64 21 ???. Last digits of the answer are 57, so we now have 6421 ? 57.

Thinking mod 11 we have 6 42 1? 57, which together has to be 3 (mod 11). $42 + 57 = 0$

(mod 11). So 6 + 1? has to be 3 (mod 11). This is only possible if instead of the question mark is a 9: 6 + 19 = 25 = 3 (mod 11). Therefore the complete answer is 6421957.

13

The number 13 is a "regular customer" in the factorisation questions during mental calculation tournaments, because, as already mentioned 7 × 11 × 13 = 1001. The second reason is that 7, 11 and 13 are all below 20, and the questions for beginners generally have 20 as the upper limit for the prime numbers.

For extracting 13th roots of very big numbers – up to 100 figures or even more – the answer often is calculated using modulo 13, since in the thirteenth power numbers have the same modulo 13 as the basic number. E.g. 2^{13} = 8192, which is 2 mod 13 just as is 2; see the table. The results are in red.

Mod 13 with basic number 2

BN	Exp.	Mod 13	BN	Exp.	Mod 13	BN	Exp.	Mod 13	BN	Exp.	Mod 13
2	1	**2**	2	5	**6**	2	9	**5**	2	13	**2**
2	2	**4**	2	6	**12**	2	10	**10**	2		
2	3	**8**	2	7	**11**	2	11	**7**	2		
2	4	**3**	2	8	**9**	2	12	**1**	2		

In the case of an even QN of 100 figures, and if from the eight figures of the answer, and the first 5 already have been found, there can be these possible answers for the last three figures: 014, 264, 514 of 764. The difference between these numbers is not a multiple of 13, so the use of modulo 13 is not a matter of "gambling".

BN 13	BN^3	BN 13	BN^3	BN 13	BN^3	BN 13	BN^4	BN 13	BN^4	BN 13	BN^4
1	1	5	8	9	1	1	1	5	1	9	9
2	8	6	8	10	10	2	3	6	9	10	3
3	1	7	5	11	5	3	3	7	9	11	3
4	12	8	5	12	12	4	9	8	1	12	1

For all clarity I quote a remark of Wim Klein: "the difficulty of a big root is not in the size of the question number but in the size of the answer number". For example he extracted the 73rd root of a 500 figure number. The answer has 7 figures. A thirteenth root of a 100 figure number gives an answer of 8 figures.

For other exponents 13 is not useful, as to be seen in the table. We see that in the third power e.g. that 1, 3 and 9 are all 1 (mod 13) because in mod 13 $1 \times 3 = 3 \times 3 = 9 \times 3$ 27 = 1 (mod 13).

In the fourth power we see the consequences of the mirror effect; 1^2 and 12^2 are 1 (mod 13); 5^2 and 8^2 are 12 (mod 13). The mirror effect appears again when we square 1 and 12 so that we now have four numbers 1 (mod 13): with 1, 5, 8 and 12.

27

This has the same nature as 9, In the third power and higher all the numbers divisible by 3 give 0(27). So 27 is in fact only useful in multiplications, divisions in "the first and second power".

As $27 \times 37 = 999 = 1000 - 1$, it is practical to work in groups of three figures from right to left, and add the "individual" moduli 27. E.g. 624187962 = 624 = 3 (mod 27) + 187 = 25 (mod 27) + 962 = 17 (mod 27), overall mod 27: 3 + 25 + 17 = 18 mod 27.

In the chapter "Diophantine equations" you'll see that with this type of question mod 27 calculation is the clue to the solution.

37

A useful number, especially for integer square roots, but not in the third roots.

Working with this number it is by far the best way to work by splitting up the given number from right to left in groups of three figures, as $27 \times 37 = 999 = 1000 - 1$. So $10^3 = 1$ (mod 37), and by consequence also 10^6, 10^9 et cetera, as long the exponents are divisible by 3. Then take the mod 37 of each group, add them and then determine the mod 37 of the whole number.

As an example we take the question number below, for extracting the integer square root:

$$\sqrt{436\ 100\ 423\ 641}$$

641 = × 10^0 = 12 mod 37; 423 = × 10^3 = 16 mod 37; 100 = × 10^6 = 26 mod 37; 436 = × 10^9 = 29 mod 37

So the mod 37 of the whole number is 12 + 16 + 26 + 29 = 83 = 9 (mod 37) so the basic number is either 3 or 34 mod 37.

Modulo 37 for the squares

BN	BN² mod 37	BN	BN² mod 37	BN	BN² mod 37
1	1	13	21	25	33
2	4	14	11	26	10
3	9	15	3	27	26
4	16	16	34	28	7
5	25	17	30	29	27
6	36	18	28	30	12
7	12	19	28	31	36
8	27	20	30	32	25
9	7	21	34	33	16
10	26	22	3	34	9
11	10	23	11	35	4
12	33	24	21	36	1

For following the workout of this question please take the table for the squares of numbers up to 1000.

Then: 435600 < 436100 < 436921, so the first three figures of the answer are 660.

For finding the last three figures these numbers are the candidates.

121²	14641	621²	385641
129²	16641	629²	395641
371²	137641	871²	758641
379²	143641	879²	772641

Have a look at the thousands of the squares: 14, 16, 758 and 772; they all are even. As the QN has odd thousands, 3641, the numbers 121, 129, 871 and 879 are not candidates, so the numbers 371, 379, 621 and 629 remain as candidates.

Again: if once a thousand is even in a square, adding more thousands leaves the even status of the thousands unimpeded according to the $(a + b)^2$ principle.

What we know now about the answer is the first three figures, 660, 31 (mod 37) and the complete answer number is either 3 or 34 mod 37. This means that the last three figures have to be either 3, to get 34 (mod 37) or 9 to obtain 3 (mod 37).

The candidates: 371 = 1 (mod37); 379 = 9 (mod 37); 621 = 29 (mod 37); 629 = 0 (mod 37). None of these candidates results in 34 of the complete answer: 31(660) + 1 = 32; 31 + 29 = 23; 31 + 0 = 31.

31 + 9 = 3 (mod37), so 379 is the correct choice, the complete answer of sqrt 436 100 423 641 is 660379.

BN	BN³	Mod 37	BN	BN³	Mod 37	BN	BN³	Mod 37
1	1	1	10	1000	1	26	17576	1
14	2744	6	29	24389	6	31	29791	6
2	8	8	20	8000	8	15	3375	8
7	343	10	33	35937	10	34	39304	10
21	9261	11	25	15625	11	28	21952	11
5	125	14	13	2197	14	19	6859	14
18	5832	23	32	32768	32	24	13824	23
9	729	26	16	4096	26	12	1728	26
3	27	27	4	64	27	30	27000	27
17	4913	29	22	10648	29	35	42875	29
6	216	31	23	12167	31	8	512	31
11	1331	36	36	46656	36	27	19683	36

Working with 37 I stumbled there: extracting an integer cubic root, the answer I thought to be the right one was 4 (mod 37) and it should be a number 3 (mod 37). After an examination I found: 1^3, 10^3 and 26^3 are all 1 (mod 37). Worked out in a table we see 12 triples of numbers who have the same mod 37.

The link is the number 10, as 1^3 and 10^3 are 1 (mod 37), $10 \times 10 = 100$ and $1 \times 1 = 1$, therefore $26^3 = 1$ (mod 37) too.

For enlightening we take another number, multiply it two times with 10 and then find the same modulo 37: $7 \times 10 = 33$ mod 37.

This makes the use of mod 37 for integer cube roots ineffective.

41

My question was: What will happen as 41 is:

- The sum of two different squares just one way one way: $5^2 + 4^2$
- A multiple of $4 + 1$
- A multiple of $5 + 1$

Look what the outcome is. About 41; we know that $41 \times 271 = 11,111 \times 9 = 99,999 = 100,000 - 1$

To summarise: In fact working with modulo 41 is not very difficult, but it is not very practical; it takes some more time, but does not reduce the quantity of work.

BN				BN^4 (41)
1	9	32	40	1
11	17	24	30	4
4	5	36	37	10
2	18	23	37	16
16	20	21	25	18
7	19	22	34	23
6	13	28	35	25
12	15	26	29	31
8	10	31	33	37
3	14	27	38	40

BN					BN^5 (41)
1	10	16	18	37	1
11	12	28	34	38	3
5	8	9	21	39	9
15	22	24	27	35	14
6	14	17	19	26	27
2	20	32	33	36	32
3	7	13	29	30	38
4	23	25	31	40	40

61

Although this number is not in the list here above, my curiosity was aroused: 41 is a multiple of 4 + 1 and 5 + 1, will 61 behave in the same way, fifteen groups of four numbers with the same modulus and have – now 12 – groups of five numbers with the same modulus in the fifth power? Yes, it does! 6, 21, 43, 54, and 59 have all the same mod 61, namely 29. The link is in the number 20, as $20^5 = 3,200,000 = 1$ (mod 61). $6 \times 20 = 59$ (mod 61) $\times 20 = 21$ (mod 61) $\times 20 = 54$ (mod 61) $\times 20 = 43$ (mod 61) $\times 20 = 6$ (mod 61).

BN				BN^4 (61)
1	11	50	60	1
8	27	34	53	9
4	17	44	47	12
13	21	40	48	13
5	6	55	56	15
2	22	39	59	16
3	28	33	58	20
7	16	45	54	22
19	26	35	42	25
9	23	38	52	34
25	30	31	36	42
14	29	32	47	47
15	18	43	46	56
10	12	49	51	57
20	24	37	41	58

BN					BN^5 (61)
1	9	20	34	58	1
8	11	28	37	38	11
12	25	42	47	57	13
5	39	45	46	48	14
10	17	29	31	35	21
6	21	43	54	59	29
2	7	18	40	55	32
26	30	32	44	51	40
13	15	16	22	56	47
4	14	19	36	49	48
23	24	33	50	53	50
3	27	41	52	60	60

In everyday practice 61 is not a practical number to work with. It is no number for modulo calculation, it owes its appearance here to the fact of being a multiple of 4 + 1 and 5 + 1.

73

As 73 × 137 = 10,001 one could work with 73 by means of + and – when subdividing a number in groups of four digits, from right to left.

As 73 is a multiple of 3 + 1 and 4 + 1, we find in the third power 24 groups of numbers with the same mod 73 and 18 groups of four numbers with the same mod 73.

I cannot remember ever having seen an article in which working with modulo 73 was mentioned.

To demonstrate how unpractical working mod 73 is, see this example.

5812 × 8973 = 52151076. The check is with mod 73:
5812 = 45 (mod 73)
8973 = 67 (mod 73)
45 × 67 = 3015 = 22 (mod 73)
1076 = × 1 = + 54 (mod 73)
5215 = × -1, as multiples of 10,000 = -32 (mod 73)
54 – 32 = 22 (mod 73) so the answer is correct.

91

You know already from modulo 13 that in the thirteenth power the numbers have the same modulo as the basic number. The same is valid for modulo 7, in fact after the sixth power the numbers have the same seven modulus, so after the twelfth power the same is seen. Therefore it is rather simple to combine these moduli to modulo 91.

As 91 represents 7 × 13 and 91 × 11 = 1001, we can use this for calculating modulo 91. This therefore attractive because the bigger the modulus, the less is the chance that in a certain group of numbers the same modulus is possible. E.g. the number (0)384 = 20 (mod 91). This means that up to 10,000 there is no other number ending on (0)384 with the same modulus. If we count 50 × 91 further we have (0)384 + 4550 = 4934. Firstly the properties of 4934 are different from (0)384, there is only one common figure, the four, so the chance of interchange can be neglected. From 0384 we can go to 100 × 91 further, then we have 9484, The "distance" between these numbers makes it very improbable that these numbers will be interchanged, moreover 0384 calculated to any power results in different final figures, so that mistakes are excluded. For the good order: for finding a correct answer modulo calculation is a very useful expedient, although one should not rely on the modulo calculation only but also combine it with other techniques, e.g. logical reasoning.

The best way of working in modulo 91 is to create groups of six figures from right to left: 274,813 is + 539 (mod 91) = 84 (mod 91). And 813,274 = -539 (mod 91) = -84 (mod 91), so (+) 7 mod 91.

For a bigger number, splitting up the number in groups of six figures from right to left one can add the moduli 91 to find to the total. This is therefore completely well-thought out because every million is 1 (mod 91): $10989 \times 91 = 999{,}999 + 1 = 1{,}000{,}000$.

E.g. 776460010919907342271653 gives us

776460 010919 907342 271653, from right to left this is 271653 = + 382 = +18 (mod 91); 907342 = -565 = -19 (mod 91); 010919 = +90 (mod 91); 776460 = -316 = -43 (mod 91) = +48 (mod 91), overall the number is 46 (mod 91).

The use of modulo 91 calculation will be described in the chapter "Very big roots" in which you'll be informed about the integer thirteenth root of numbers of one hundred figures. For other operations there are enough other moduli available.

For modulo calculation one has the possibility to take a small prime number, which is very easy to handle, e.g. 3. 11 is much better and practical to "handle", as $9 \times 11 = 99$, and $99 + 1 = 100$. After splitting numbers into groups of 2 figures from right to left it is easy to handle. Besides which the accuracy increases, the error rate is 1:11 = 9.09%, which is a lot better. 101 is still better: 1:101 is less than 1%! But this is theory.

101

It is a matter to get accustomed to, but 101 has a lot of advantages. Let's see: Firstly, the four digit numbers: A 2 digit number multiplied by 101 gives two symmetric parts; $47 \times 101 = 4747$. If the second part is bigger, then we have a positive mod 101: 4783 = +36 (mod 101); 4711 = -36 (mod 101) or +65 (mod 101). From right to left, with digits subdivided into groups of four digits we can count with +, -, +, - et cetera. Let's see with 2231440831420641. For mod 101 we have 2231 4408 3142 0641. From right to left we have here: +35; +11; -36; +9, overall 19 (mod 101) or -82 (mod 101). Using 101 for multiplications is not difficult, for divisions it is very difficult, integer fourth roots are worked out later on. For integer fifth roots 101 behaves like 41 and 61: there are 20 groups of five numbers with the same mod 101 in the fifth power: 11, 15, 35, 48 and 93 to the fifth power all are mod 57 (mod 101). The link is in 36: $11 \times 36 = 93$ (mod 101), $\times 36 = 15$ (mod 101) $\times 36 = 35$ (mod 101) $\times 36 = 48$ (mod 101) $\times 36 = 11$ (mod 101).

Working in groups of four figures we may conclude the exponents of 10 which are not divisible by 4 result in a - mod 101, the exponents which are divisible by 10^4, 10^8 et cetera result in a positive mod 101. E.g. 3748: 48 represents the units and is positive, 37 represents the hundreds – 101-1 – and is negative. So the whole number 3748 = 48 – 37 = +11 (mod 101).

13.2.12 Multiplications

Modulo calculations can be used for every kind of operation. It is questionable if mod 101 has advantages over mod 11. 4691 × 5393 = 25 29 85 63. 4691 = 45 (mod 101), 5393 = 40 (mod 101). 45 × 40 = 1800 = -18 or 83 (mod 101). 63 and 29 are +, total 92. 25 and 85 are negative, total -110, = -9. Since 92 – 9 = 83, the conclusion is that the answer is correct.

13.2.13 Divisions

For divisions mod 101 is much more difficult, as we will soon see.

Just for argument's sake; two integer divisions, randomly found in a test paper.

$$1\ 6250\ 2912 \div 32\ 89$$

The answer has 5 digits. The two last ones of it, 08, the very first one is 4 as 4 × 3289 = 13156 and 5 × 3289 = 1 6445. The answer so far is 4 xx 08. The dividend = +1, -12 and -17 (mod 101), so the whole dividend is -28 = 73 (mod 101). 3289 = + 57 (mod 101). 73 is not divisible by 57, but by adding 73 + 10 × 101 we get 1083 ÷ 57 = 19, so the complete answer has to be 19 (mod 101). 4 xx 08 = 12 (mod 101), and so to complete the answer we need 7 (mod 101) which agrees with 9400, so the complete answer is 49408.

$$554\ 5401\ 1916 \div 79\ 13\ 41$$

The answer will have 5 digits, of which the first one will be 7 as 7 × 7913 = 55391. The two last ones will be 76, as 41 × 76 = 3116. The answer so far 7 xx 76. The dividend (554) 49 – (5401) 53 – (1916) 3 = -7 = 94 (mod 101). -7 = 94 (mod 101). The divisor: 79 – 13 + 41 = 107 = 6 (mod 101). Because 94 is not divisible by 6 we add 4 × 101 to get 498 ÷ 6 = 83. 7 xx 76 = 83 (mod 101), so the missing figures are 00, so the final answer will be 70076.

In fact, as an expedient in divisions the number 101 is not practical. For finding e.g. the solution of 19 (mod 101) ÷ 83 a lot of time-consuming work needs to be done. Using the algorithm of Euclid is only a theoretic possibility, in practice in mental calculation it is unfortunately ineffective. In this case after adding 22 × 101 we get 2241 ÷ 83 = 27.

13.2.14 Integer fourth roots mod 101

In chapter 15 extracting integer fourth roots mod 101 is worked out more elaborate.

13.2.15 The fifth power

In the fifth power there are 20 groups of five numbers with the same modulus. The linking number is 36. 1 × 36 = 36 × 36 = 84 × 36 = 95 × 36 = 87, all modulo 101. 101 is a multiple of 4 + 1 and of 5 + 1. Also here we see that in the fourth power we have groups of four numbers with the same modulus and in the fifth power groups of five numbers with the same modulus.

BN					BN⁵ (mod101)
1	36	84	87	95	1
22	30	70	85	96	6
10	32	41	57	62	10
4	33	43	45	77	14
13	20	23	64	82	17

During an algebra lecture the professor asked the students: "what is the modulo 101 of 2^{50}?". "Professor, may I give it a try?" Yes. My answer: 100, his answer -1. There is no difference. How did I do it? Well $2^{10} = 1024 = 14$ (mod 101), and by consequence $2^{20} = 95$ (mod 101). To square this we can also take the "mirror" number 6, which square is 36, so now we have the mod 101 of 2^{40}. If we now multiply 36 × 14 we have the mod 101 of 2^{50}. The result is 504 which is easily reduced to 100 (mod 101) or, if you want, -1 (mod 101).

137

$137 = 11^2 + 4^2$, only one time the sum of two different squares, so it is a prime number. Ok, 137 × 73 = 10.001, which raises the question if the number could be an expedient for modulo calculation. Unfortunately, it is not. In the fourth power we see that there are 34 groups of four digit numbers, which calculated to the fourth power with the same mod 137. E.g.: 31, 51, 86 and 106 all calculated to the fourth power have 4 mod 137.

Just for an example one number is worked out:

3843, 4907, 5093, 6157

$6157 \times 10^0 = 6157 \times 1 = 6157 \div 137$ gives remainder – 8 as 45 × 137 = 6165

$5093 \times 10^4 = 5093 \times -1$, 5093 ÷ 137 gives remainder 24, as 37 × 137 = 5069. As 5093 is $\times 10^4$ the remainder is – 24

$4907 = \times 10^8$, and 9999 × 10001 = 99999999, so 10^8 has remainder 1 so 4907 ÷ 137 = -25 mod 137 as 36 × 137 = 4932

3843 is $\times 10^{12} \times -1$ as $10^4 \times 10^8 = -1 \times +1$ gives -1. 3843 ÷ 137 gives remainder 7 as 28 × 137 = 3836. In this case – 7.

Finally we add: -8 – 24 – 25 – 7 = -64 mod 137, or + 73

181

$181 = 10^2 + 9^2$. And again we recognise some elements. Now we have seen three prime numbers which calculated to the fifth power have the following element in common; there are groups of five numbers with the same modulus, and these numbers are all the sum of two different squares and a multiple of $10 + 1$. This evokes the question: is this always the same? Therefore 181 was examined, as it is a multiple of $3 + 1$, $4 + 1$ and $5 + 1$. For real calculation 181 is neither needed nor practical, it is a matter of "I want to know". And indeed, 181 behaves in the same way.

There are according to the expectations 60 groups of three numbers which calculated to the third power result in 1 (mod 181): 1, 48 and 132 → 48 × 132 = 6336 = 1 (mod 181). There are 45 groups of four numbers which calculated to the fourth power result in 1 (mod 181). They are 1, 19, 162, 180.

Finally 10, 47, 58, 83 and 164 calculated to the fifth power give all mod 88 (mod 181). The link is 42: 10 × 42 = 58 (mod 181) × 42 = 83 (mod 181) × 42 = 47 (mod 181) × 42 = 164 (mod 181) × 42 = 10 (mod 181).

For the interested people: 1 ÷ 181 = 0.005524861878453038674033149171270718232044198895027624309392265193370 165745856353591160220994475138121546961325966685082872928176795580110497237569060773480662983425414364640883977 9, a number of 180 digits. The cut is made after the 90th digit and if you start with 005524 et cetera and put right under this the second part, beginning with 994475 et cetera you'll get a number consisting of 90 consecutive nines.

1001

When modulo calculating it is always practical to take a number which itself or a multiple thereof lies at 10^n. Think back to 11 with 100; 27 and 37 with 1000, in theory 41 with 100,000 and 73 and 137 with 10,000.

The number 1001 works more practically, moreover this because it is the product of $7 \times 11 \times 13$. If numbers are divided in groups of three figures, this works easily.

If a number represents 10^3 or another odd 10^3 group, the numbers are to be counted as the units, in other cases the numbers are to be counted as − 1001, of which neither action is very difficult.

We consider the number 436 100 423 641, which we saw before. We go from right to left.

   641, units, so positive
   423, thousands, so negative

100, millions, so positive

436, billions, so negative.

The complete number: 641 − 423 + 100 − 436 = rest − 118 (mod 1001).

Very simple is a number like 246246, immediately to be recognised as 246 × 1001. And after some practice you'll recognise too a number like 6,522,516. Subtract 6,522 − 516 = 6,006 and you see that the number is divisible by 1001.

13.3 Answers

Answers for the nines test:

68 × 97 = 6596 = 8 (mod 9). 68 = 6 + 8 = 14 = 5 (mod 9); 97 = 7 (mod 9). 5 × 7 = 35 = 8 (mod 9).

374 × 821 = 307054 = 1 (mod 9). 3 + 7 + 4 = 5 (mod 9); 8 + 2 + 1 = 2 (mod 9). 5 × 2 = 10 = 1 (mod 9).

1637 × 8251 = 13506887. 1 + 6 + 3 + 7 = 8 (mod 9); 8 + 2 + 5 + 1 = 7 (mod 9). 8 × 7 = 56 = 2 (mod 9) The answer: 1350 = 0 (mod 9); 6 + 8 + 8 + 7 = 29 = 2 (mod 9). The answer passes the test.

Answers for the elevens test:

84 × 91 = 7644 = 10 (mod 11). 44 can be ignored because it is 0 (mod 11), and 76 = 10 (mod 11). 84 = 7 (mod 11), 91 = 3 (mod 11) and 7 × 3 = 10 (mod 11), so the answer passes the test.

397 × 643 = 255271. 71 = 5 (mod 11), 25 + 52 = 0 (mod 11), so the answer is 5 (mod 11). 3 + 97 = 100 = 1 (mod 11), 6 + 43 = 49 = 5 (mod 11), 1 × 5 = 5 (mod 11). The answer passes the test.

4653 × 9101 = 42346953. 53 = 9 (mod 11), 69 = 3 (mod 11), 34 = 1 (mod 11), 42 = 9 (mod 11), the sum = 22 = 0 (mod 11). 46 + 53 = 99 = 0 (mod 11) × the modulus of 9101 is of no importance as 0 times any number remains 0. The answer passes the test.

8633 ÷ 97 = 89. With the dividend 8633, the 33 can be ignored, 86 = 9 (mod 11). 97 = 99 − 2, so 9 (mod 11). 9 ÷ 9 = 1, the answer will be 1 (mod 11). Then a number ending on 3 divided by a number ending on 7 results in 9 uniquenely. Which two digit number ending in 9 is 1 (mod 11)? There is only one possibility, this is 89.

124307 ÷ 197. 12 + 43 = 0 (mod 11), 07 = 7 (mod 11), the dividend is 7 (mod 11). The divider 1 + 97 = 10 (mod 11). In modulo 11 calculation 7 ÷ 10 = 4, which is the mod 11 of the answer. About the answer we can say that the last digit is 1, as 7 ÷ 7 = 1. The first digit of the answer can be found by reliable estimation: 6 × 197 is a bit less than 1200, 7 × 197 is a bit less than 1400, so 6 is correct. Now we have 6?1 to be 4 (mod 11). 6 01 =

7 (mod 11). Per ten more there is a reduction of 1 mod 11. 7 − 3 = 4 so 631 will be the correct answer. Check: 6 + 31 = 37 = 4 (mod 11).

35336864 ÷ 4276. 64 + 68 = 0 (mod 11), 35 33 is 2 (mod 11). 76 + 42 = 8 (mod 11). In modulo 11 calculation 2 ÷ 8 = 3 (mod 11), the modulus of the answer. The first figure of the answer will be 8, as 8 × 42 + is more than 340. The last ones are nasty: 64 ÷ 76 have as candidates 14, 39, 64 and 89. Now a new element is introduced: how many twos are in the dividend and the divider? In 336 864 there are 2^5. In 4276 there are 2^2. $2^5 \div 2^2 = 2^3$. This means that the answers 14, 39 and 89 are excluded. So 64 remains. The provisional answer then is 8? 64. 80 + 64 = 1 (mod 11). To obtain 3 mod 11 we have to add 200, then we have 8264, which is 3 (mod 11).

13.4 Summary

The reader will conclude that of all the mentioned moduli only a small part is used in the all-day practice. Indeed, that is right!

But studying the various moduli was very instructive to me and it enriched my Insight in the magic number world enormously. For those who wonder what is the use of all this, I have only one answer, when I quote George Lane: "We should never forget that all the operations we do are interlinked on one or another way".

For myself modulo calculation is an indispensable expedient in a wide variety of operations.

14 Powering

Simply said, powering is multiplying a given number by itself, 2 times we have a square, we can take this as many times as we want. And then we enter the magic world of the structures in the numbers. The number in superscript is the exponent, which indicates how many times the given number was multiplied by itself.

What happens? Studying what happens when powering offered me a sea of surprises as later on you will see too.

Other than with the number 1 we see three things:

- The numbers are getting bigger and bigger – the quantitative aspect
- And the last digits are changing – the qualitative aspect
- And if we continue long enough, we see the basic number return, this is all according to a predictable structure

Table 1 hereunder gives insight into the changes of the last two digits:

Exp.1	Exp. 2	Exp. 4	Exp. 5	Exp. 10	Exp. 20
01, 25, 76	24, 49, 51, 74, 75, 99	07, 18, 32, 43, 57, 82, 93	6, 16, 21, 36, 41, 56, 61, 66, 81, 86, 96	04, 09, 11, 14, 19, 29, 31, 34, 39, 44, 59, 64, 69, 79, 84, 89, 91, 94	The rest

Table 1 indicates after which power a number returns in its basic shape. 01, 25 and 76 never change, with e.g. 51 tumbles, we get 51, 01, 51.

A lot of numbers return after the twentieth power, i.e. in the case of the last two figures. So 13, a number in the category 20, returns as 13 in the 21st power. If we want to get 013 back we have to calculate to the 100th power. We will not go so far! The even numbers, only divisible by 2, change in this way: 02 gets 52; 14 gets 64 when powering.

Table 2

BN	Exp.	2 LD	BN	Exp.	2 LD	BN	Exp.	2 LD	BN	Exp.	2 LD
13	1	**13**	13	11	**37**	29	1	**29**	64	1	**64**
13	2	**69**	13	12	**81**	29	2	**41**	64	2	**96**
13	3	**97**	13	13	**53**	29	3	**89**	64	3	**44**
13	4	**61**	13	14	**89**	29	4	**81**	64	4	**16**
13	5	**93**	13	15	**57**	29	5	**49**	64	5	**24**
13	6	**9**	13	16	**41**	29	6	**21**	64	6	**36**
13	7	**17**	13	17	**33**	29	7	**9**	64	7	**4**
13	8	**21**	13	18	**29**	29	8	**61**	64	8	**56**
13	9	**73**	13	19	**77**	29	9	**69**	64	9	**84**
13	10	**49**	13	20	**01**	29	10	**1**	64	10	**76**
			13	21	**13**	29	11	**29**	64	11	**64**

BN = basic number   Exp. = exponent   2 LD = 2 last digits

We see how the number 13 changes its last figures: 13^{20} ends in 01 and by consequence 13^{21} ends in 13. In the same way, 29^{10} ends in 01 and 29^{11} ends just as 29^1 on 29. And 64^{11} ends again on 64.

As it is impossible to discuss all the numbers in all the imaginable powers we'll restrict to the fifth power.

14.1 The squares

Table 3

BN	BN²
1	1
2	4
3	9
4	16
5	25
6	36
7	49
8	64
9	81

1 + 9	10
2 + 8	10
3 + 7	10
4 + 6	10

Table 3 gives the basic structure in the squares. Quickly visible is that we have here several times the same final digits and that the sum of the basic numbers with the same final digits when squared, still is 10. What we see more is that the 1, 5, 6 remain unchanged, that in the squares the figures 2, 3, 7 and 8 are missing. This This evokes the question: is there more to be seen? Yes!

Here we see two examples of squared 2 digit numbers and we cannot miss: if the final digits are the same, the structure is that the basic numbers are linked: either together 100, together 50 or differ 50.

Where does this all come from? From the algebra, the variant of the famous algebra theorem of $(a+b)^2 = a^2 + 2ab + b^2$. $7^2 + (86 \times 100) = 8649 = 93^2$.

Table 4

1	01	1 + 99	100
49	2401	49 + 51	100
51	2601	1 + 50	51
99	9801	49 + 50	99
7	49	7 + 93	100
43	1849	43 + 57	100
57	3249	7 + 50	57
93	8649	43+ 50	93

What we see more: the squares are either 0 (mod 4) for the even numbers, or the odd numbers when squared give without exception 1 (mod 4). We can safely conclude: numbers which are either 2 (mod 4) or 3 (mod 4) can never be a square.

BN	BN²
7²	049
243²	59049
257²	66049
493²	243049
507²	257049
743²	552049
757²	573049
993²	986049

7 + 993	1000
243 + 757	1000
257 + 743	1000
493 + 507	1000

Table 5. The above mentioned leads us to the following question: does something likely also exist in bigger numbers? Yes; have a look at table 5.

For qualities as equal final digits this is valid: From 3 digits on there can be found 8 numbers with the same final digits, related to the basic number. E.g. if you want to have the row of numbers whose squares end on 8649, then here they are: 93, 1157, 3843, 4907, 5093, 6157, 8843, 9907.

More is to be said about this row. 8649 is 9 (mod 16), so the basic numbers may be 3, 5, 11 or 13 (mod 16). For all clarity here is the row inclusive the 16 moduli between brackets: 93 = 13 (mod 16), 1157 = 5 (mod 16), 3843 = 3(mod 16), 4907 11 (mod 16), 5093 = 5 (mod 16), 6157 = 13 (mod 16), 8843 = 11 (mod 16), 9907 = 3 (mod 16). To make it easier to compose rows of numbers like this it is advisable to invoke mod 16 calculation.

In the case of even numbers it is a little bit different.

BN	BN²	BN	BN²
122	14884	46	2116
128	16384	204	41616
372	138384	296	87616
378	142884	454	206116
622	386884	546	298116
628	394384	704	495616
872	760384	796	633616
878	770884	954	910116

Table 6. Here we see a difference of a multiple of 500 between the squares of the basic numbers 2, 6, 10 or 14 (mod 16) and the numbers 0, 4, 8 or 12(mod 16). E.g. 128² − 122² = 1500. If you want a series of four digit numbers for which the squares end in 2884, here they are: 378 (10 mod 16), 2122 = (10 mod 16), 2878 = (14 mod 16), 4622 = (14 mod 16), 5378 = (2 mod 16), 7122 = (2 mod 16), 7878 = (6 mod 16), 9622 = (6 mod 16).

16² = 2 56, the hundreds are 2 (mod 4). 216² = 4 66 56, the hundreds 66 are also 2 (mod 4).

Table 7

Final digit	Tens	(H)undreds (E)ven / (O)dd
1	01	Always even
	21	Always odd
	41	Always even
	61	Always odd
	81	Always even
4	04	E with H 0 (mod 4) and O with H 3 (mod 4)
	24	E with H 2 (mod 4) and O with H 3 (mod 4)
	44	E with H 6 and 14 (mod 16) and O with H 1 and 13 (mod 16)
	64	E with H 0 and 4 (mod 16) and O with H 1 and 9 (mod 16)
	84	E with H 4 and 12 (mod 16) and O with H 3 and 7 (mod 16)
5	25	Always even, H never 4 of 8
6	16	E with H 0 and 12 (mod 16), O with H 5 and 13 (mod 16)
	36	E with H 0 and 8 (mod 16) and O with H 3 and 15 (mod 16)
	56	E with H 2 and 6 (mod 16) and O with H 3 and 11 (mod 16)
	76	E with H 6 and 14 (mod 16) and O with H 5 and 9 (mod 16)
	96	E with H 8 and 12 (mod 16) and O with H 1 and 9 (mod 16)
9	09	Always even , H all E: 0,2,4,6,8,10,12,14 (mod 16)
	29	Always odd, H all O: 1,3,5,7,9,11,13,15 (mod 16)
	49	Always even H all E: 0,2,4,6,8,10,12,14 (mod 16)
	69	Always odd , H all O: 1,3,5,7,9,11,13,15 (mod 16)
	89	Always even H, all E: 0,2,4,6,8,10,12,14 (mod 16)

Table 7 gives insight into the possibilities of the numbers which *can* be a square. It is recommended to take a table with at least all the squares up to 100 to verify what is mentioned in this table. As the hundreds are a multitude of 4 (mod 16) the hundreds are related to mod 16.

For the odd numbers it is seen that the hundreds increase by an even number. E.g. 101^2 = 10 2 01, 201^2 = 40 4 01, as the sum of two odd numbers is an even number.

For even numbers the sum of two of them is divisible by 4, so that the increase of the hundreds is divisible by 4. 12^2 = 1 44, the hundred is 1 (mod 4); 312^2 = 9 73 44. Here the hundreds are 73, again 1 (mod 4).

14.2 Creating squares

This table enables us to make squares with the qualities we want, e.g. a number ending on 29 and hundreds 65, so final digits 6529. We know $23^2 = 529$; we look for 6000. Either we use the cross method or we take $(a + b)^2$, $23 + 23 = 46$, here we have the six we want and $1023^2 = 1,046,529$. To complete the famous 8: 1023, 2273, 2727, 3977, 6023, 7273, 7727, 8977. Examine this yourself: $6529 \equiv 1 \pmod{16}$ so the basic numbers will be 1, 7, 9 or 15 (mod 16). It is an interesting matter and I can only recommend you to create a lot of squares to gain confidence with it.

Table 8

1023	6023	1023 + 8977 = 10,000
2273	7273	2273 + 7727 = 10,000
2727	7727	2727 + 7273 = 10,000
3977	8977	3977 + 6023 = 10,000

Let's say we want a square ending in 4969 with the basic number 13. $13^2 = 169$, so we want to add 4800. $2 \times 13 = 26$ and $48 \div 26 \equiv 48$ or 98, we take 4813, and indeed $4813^2 = (2316)\,4969$. We could also take as basic number $63^2 = 3969$. So now we have to add 1000. $2 \times 63 = (1)26$ and $1000 \div 26 \equiv 35$ or 85, and indeed $3563^2 = (1269)\,4969$. To complete the series: $5000 - 4813 = 187$, $5000 + 187 = 5187$, $10000 - 187 = 9813$. We get in order of magnitude: 187, 1437, 3563, 4813, 5187, 6437, 8563, 9813.

Also helpful is modulo 16 calculation. 969 ET (even thousand) is 9 (mod 16) so the BN (basic number) will be 3, 5, 11 or 13 (mod 16). We can also find all the squares ending in 34969. We get in order of magnitude: 187, 6437, 43563, 49813, 50187, 56437, 93563, 99813.

Generally from the squares up to thousand and higher one can say: if you want to have a number of equal final figures in the squares, there are always as many possibilities as the power of 10. So for 6 equal final figures you can find up to 1,000,000 eight squares.

If you want to have 8 squares ending on 687321, then start with $139^2 = 19321$. We now miss 668000. We divide ?668 ÷ 139 and get 12, so the first square we look for is 12 ÷ 2 = 6, so $6139^2 = (37)687321$. We now can easily find: 993861, 493861 and 506139. $687321 = 25 \bmod 64$, which means that the basic numbers can be 5, 27, 37 or 59 mod 64. The ones we have are 6139 and $87611 = 59 \bmod 64$ and their complements to 1,000,000, 912,389 and 993,861 are by consequence 5 mod 64. Now 500,000 minus 6139 = 493861, 37 mod 64. Then we get another 37 mod 64: 500,000 − 87611 = 412389. Now 1,000,000 − 412,389 = 587,611 and we have them all, in sequence: 6139, 87611, 412389. 493861, 506139, 587611, 912389 and finally 993861. The choice for mod 64 is because this is 2^6 as 1,000,000 is 10^6.

14.3 The fives

Squaring numbers ending with a 5 means that the two final digits unchangeably will be 25, the hundreds will never be odd, and never be 4 or 8, and if the basic number (BN) is either 25 or 75 the hundreds will unchangeably be 06 or 56.

BN	BN²	BN	BN²	BN	BN²	BN	BN²
35	1225	45	2025	25	625	125	15625
215	46225	205	42025	225	50625	375	140625
285	81225	295	87025	275	75625	625	390625
465	216225	455	207025	475	225625	875	765625
535	286225	545	297025	525	275625		
715	511225	705	497025	725	525625		
785	616225	795	632025	775	600625		
965	931225	955	912025	975	950625		

Numbers 15, 35, 65 and 85 will give squares ending on 225. Numbers 05, 45, 55 and 95; their squares will end on 025.

Concerning the squares of the non 25 or 75 group we see that the thousands differ by a multiple of 5000. Squares of 25 or 75 numbers differ by multiples of 25000.

An anecdote

As a boy of 9 years I knew all the 2 × 2 multiplications by heart. Being 11 years old I was taught by a teacher a trick with squaring numbers ending on 5. It is this one: If you want to square 25, you take 3 × 2 and put 25 behind, and indeed, $25^2 = 625$. And for 55^2 you take 6 × 5 and put 25 behind, indeed, $55^2 = 3025$. After having understood the nature of the trick I asked the teacher: do you know how much is 865^2? He did not know, and he could not know, while not commending the 2 × 2 multiplications by heart. Well, the principle is the same: 86 + 1 = 87 × 86 = 7482, so $865^2 = 748225$. His reaction: Ohh, I cannot do that……

14.3.1 Working with "the trick"

Ok, we now know that we can find something about the nature of a 5 square by multiplying 2 consecutive numbers.

Summarized: numbers 15, 35, 65 and 85 will result in squares 225, 25 and 75 will result in squares 625, numbers 05, 45, 55 and 95 will result in squares 025. And as it is with the tens, it is with the hundreds:

- Numbers 05, their hundreds increase with 10, their thousands with 1: $5^2 =$ (00) 25, $105^2 = 1\ \mathbf{10}\ 25$
- Numbers 15 their hundreds increase with 30, their thousands with 3: $15^2 =$ $\mathbf{2}\ 25$, $115^2 = 1\ \mathbf{32}\ 25$
- Numbers 25, their hundreds increase with 50, their thousands with 5: $15^2 =$ $\mathbf{6}\ 25$, $125^2 = 1\ \mathbf{56}\ 25$
- Numbers 35, their hundreds increase with 70, their thousands with 7: $35^2 =$ $\mathbf{12}\ 25$, $135^2 = 1\ \mathbf{82}\ 25$
- Numbers 45, their hundreds increase with 90, their thousands with 9: $45^2 =$ $\mathbf{20}\ 25$, $145^2 = 2\ \mathbf{10}\ 25$
- Numbers 55, their hundreds increase with 110, their thousands with 1: $55^2 = \mathbf{30}\ 25$, $155^2 = 2\ \mathbf{40}\ 25$
- Numbers 65, their hundreds increase with 30, their thousands with 3: $65\ 2 = \mathbf{42}\ 25$, $165^2 = 2\ \mathbf{72}\ 25$
- Numbers 75, their hundreds increase with 50, their thousands with 5: $75^2 = \mathbf{56}\ 25$, $175^2 = 3\ \mathbf{06}\ 25$
- Numbers 85, their hundreds increase with 70, their thousands with 7: $85^2 = \mathbf{72}\ 25$, $185^2 = 3\ \mathbf{42}\ 25$
- Numbers 95, their hundreds increase with 90, their thousands with 9: $\mathbf{90}\ 25$, $195^2 = 3\ \mathbf{80}\ 25$

Let's say we look for a square of a 5 number, ending on 832 25. So the question is: which two numbers multiplied end on 832 ? $11 \times 12 = 132$, and the increase of the hundreds is 3 ($111 \times 112 = 12\ \mathbf{4\ 32}$).

$832 - 132 = 700 \div 3 \equiv 9$, so we have to take $911 \times 912 = 830\ \underline{\mathbf{832}}$. So 9115^2 meets the requirements: $830\ \underline{83225}$. This number is 9 mod 16, so the basic number will be 3, 5, 11 or 13 (mod 16). Finally we find this: 885 as $10,000 - 9,115 = 885$. $885 = 5$ mod 16, $9115 = 11$ mod 16 then the other numbers will be 5 and 13 mod 16 and we find 4635, 11 (mod 16) and 5365, 5 mod 16.

Question: find 4 squares ending on 80025.

Solution: the numbers 05, 45, 55 and 95 result in squares ending on 025, and we need even 80025. If we take 105 the thousands increase by 1. As we need 80,000 we take 800×1, so 8005, which squared is 64080025. As we work in the area of 5 final digits, we take $100,000 - 8005 = 91,995^2 = 8463080025$. The other ones are any then thousand + 1995 and any ten thousand + 8005.

14.4 The Cubes

Table 1		Table 2		Table 3		Table 4		Table 5	
BN	BN³	BN	BN³	BN	BN³	BN	BN³	BN	BN³
1	1	1	1	2	8	3	27	4	64
2	8	11	1331	12	1728	13	2197	14	2744
3	27	21	9261	22	10648	23	12167	24	13824
4	64	31	29791	32	32768	33	35937	34	39304
5	125	41	68921	42	74088	43	79507	44	85184
6	216	51	132651	52	140608	53	148877	54	157464
7	343	61	226981	62	238328	63	250047	64	262144
8	512	71	357911	72	373248	73	389017	74	405224
9	729	81	531441	82	551368	83	571787	84	592704
		91	753571	92	778688	93	804357	94	830584

Table 1 gives us a basic insight into what happens when cubing. We have a look at the last one digit. 1, 4, 5, 6 and 9 remain unchanged, 2 and 8 interchange, just as 3 and 7, 7 and 3 and 8 and 2. He/she who knows this little row can do integer cube roots of at least 6 digits. E.g. cube root of 658503. As 512 < 658 < 729 the first digit of the answer is 8. The last digit of the question is 3 so the last digit of the answer is 7, the final answer is 87.

From Table 2, the basic numbers ending on 1. We look at the last two digits of the cubes: 1, 31, 61, 91 et cetera There is an increase of 3 in the tens. This is based on what is called "the iteration of Newton". Based on the formula $(a + b)^3$ we get $a^3 + 3a^2b + 3ab^2 + b^3$. We take for the tens $a = 10$ and for the units $b = 1$. For 11^3 the working out is $10^3 = 1000$; $3a^2b = 3 \times 100 \times 1 = 300$; $3ab^2 = 3 \times 10 \times 1 = 30$; $b^3 = 1$, so in total 1331.

This formula explains the increase, henceforth called the 'jump' of the tens.

In table 3, with basic numbers ending on 2 we see an increase in the tens of 2. Table 4, basic numbers ending on 3 shows us an increase in the tens of 7. Table 5, basic numbers ending on 4 shows us an increase in the tens of 8.

The reason for this is to be found in the iteration of Newton. According to this the iteration of the tens is submitted to this formula: $3 \times 1^2 \times 10 = 30$ for the numbers ending on 1. For numbers ending on 2 the increase of the tens is $3 \times 2^2 \times 10 = 20$. For numbers ending on 3: $3 \times 3^3 \times 10 = 70$ et cetera et cetera We will also see a symmetry: for numbers ending on 6 is valid $3 \times 6^2 \times 10 = 80$, for the 7 ending numbers the increase of the tens is 70 and for 8; $3 \times 8^2 \times 10 = 20$.

As there is always a strict logic in the number world it lies at hand that the iteration of Newton does not only work in tens, but also in hundreds and thousands and whatever more we want. Let's have a look at table 6. We see that indeed according to Newton the hundreds increase with 300: $3 \times 1^2 \times 100 = 300$. From now this increase will be named as the "Jump". For clarity see table 7. Here the jump is $3 \times 3^2 \times 100 = 2700$. With knowledge we can now calculate (e.g.) which ones are the last four digits of 4197^3. 97^3 = 912673. The jump for cubing 97 is $3 \times 97^2 \times 100 = 2700$ per hundred. $41 \times 2700 =$ (mod 11) 0700, so the cube of 4197 should end on 2673 + 0700 = 3373. And indeed, 4197^3 = 7392935 3373.

And 23097^3? We now take a = 1000 and get $3 \times 97^2 \times 1000 = 27000$. And $23 \times 27000 =$ (6) 21000. So 23097^3 should end on 12673 + 21000 = 33673. And indeed, no surprise today; 23097^3 = 123215891 33673.

14.5 Powering, Q&A

3 digit BN, their eight squares must end on 969.
We can start with $13^2 = 169$. We miss 800. As 3 + 3 = 6 (2ab) we know that for each 100 there is an increase of 600; so we have to add 3×100. Indeed $313^2 = 97\underline{969}$. 313 – 250 = 63^2 = 3 969. 250 – 63 = 187^2 = 34 969. 500 – 63 = 437^2 = 190 969. As we have now 63, 187, 313, and 437 we can simply find the rest by adding 500 to the numbers we have and then get 563, 687, 813 and 937. If we want to know which squares has an even thousand and which one an odd thousand: ET (Even Thousand) 969 = 9 (mod 16) so the basic numbers have to be 3, 5, 11 or 13 (mod 16). ET: 187 (11mod 16), 437 = 5 (mod16), 563, 3 (mod16), 813, 13 (mod16). OT (Odd Thousand) 969 are 1 (mod 16) so their basic numbers have to be 1, 7, 9 or 15 (mod 16). OT 63 (15 mod16), 313 (9 mod16), 687 (15 mod16), 937 (9 mod16). Their squares in order of magnitude are 3 969; 34 969; 97 969; 190 969; 316 969; 471 969; 660 969 and 877 969.

3 digit BN, their squares must end on 144 or 644.
This means we have numbers 0 (mod 4) and 2 (mod 4). As we know $12^2 = 144$, we have the first one. Here we can also see that the 0 (mod 4) numbers will have odd hundreds when squared. Furthermore 250 – 12, 238, 250 + 12 = 262 and 500 – 12 = 488. So 2 = 144; 238^2= 56 644, 262^2 = 68 644; 488^2 = 238144; 512^2 = 262144, 738^2 = 544644; 762^2 = 580644 and 988^2 = 976144.

8 × 4 digit BN, their squares must end on 8176.
The smallest square is $24^2 = 576$. We miss 8176 – 576 = 7600; to divide by $2 \times 24 = 48$ = 12. Our first number 1224. Immediately; 1224, 3776, 6224, 8776. As 2500 = 4(mod 16) we can do 2500 – 1224 = 1276, + 2500 = 3776. And 1224 + 2500 = 3724, + 5000 = 8724. In sequence: 1224^2 = 149 8176; 1276^2 = 162 8176, 3724^2 = 1386 8176; 3776^2 = 1425 8176; 6224^2 = 3873 8176; 6276^2 = 3938 8176; 8724^2 = 7610 8176; 8776^2 = 7701 8176.

Find some squares ending on 81225.

Easily we find $35^2 = 1225$. We miss 80000.

The following numbers – up to 100 – give a result ending on 12:

03	×	04	12
28	×	29	812
71	×	72	5112
96	×	97	9312

Here we see that 28×29 results in 812, so $285^2 = 81{,}225$. $03 \times 04 = 12$ and adding one hundred means that the hundreds increase with 700. Then $403 \times 404 = 16\,2812$, here we have our second one.

And indeed $4035^2 = 162\,81225$. Next we take 10,000 minus 285 and 4035, result respectively 9715 and 5965, there squares are 943 81225 and 355 81225.

14.6 Structure in the fourth power

We know that when we consider the last three digits in the squares we can make a list of 8 numbers with the same last three digits when squared. E.g. 013^2, 237^2, 263^2, 487^2, 513^2, 737^2, 763^2, 987^2 all end on 169. How will that be in the fourth powers? When we square a number ending on 169 the answer will end on 561. And 561 can also be obtained when we square numbers like these ones: 081, 169, 331, 419, 581, 669, 831 and 919. So for the fourth powers of numbers ending on 561 we have 16 candidates. Of course this assertion is valid for all groups of the three last digits which are a part of a square.

We may now conclude that working with three equal final digits for every fourth power number there are 16 possibilities.

To find the last digits of a fourth power we take simply the last digits of a square and square that number. $13^2 = 169$ and the last digits of 13^4 are (in case of 3 final digits) 561. What is the shape of a 4th power number?

We start with the <u>even numbers</u>. Even numbers have at least 2^1, so their squares will have 2^2 and an even number to the fourth power will have at least 2^4, so it is divisible by 16. The first one is 16. The next number which has at least 2^4 is 96. The difference is 80, which is $2^4 \times 5$. After that comes 176. So we may conclude that the last three digits of even number differ by 80, and we can make a table like 16, 96, 176, 256 et cetera et cetera Except for the even numbers ending on 0, all the even numbers calculated to the fourth power end on 16, increased by a multiple of 80.

The <u>odd numbers</u>. We start with 1^4 and the fourth power of the next odd number is $3^4 =$

81. The table goes as follows: 81,161, 241, et cetera et cetera So the shape of an odd integer fourth power is $1 + (2^4 \times 5)$. Except the numbers ending on 5, all the odd numbers calculated to the fourth power end on 1 increased by a multitude of 80.

This reduces the possibilities considerably; on the other hand there are for each number no fewer than 16 possibilities, in the case of 3 final figures.

14.6.1 The jumps

The increase of the hundreds – the "jump" – follows of course Newton's iteration. So: 13^4 ends on 8561. The jump for $13^4 = 4 \times 13^3 \times 100 = 8800$ per hundred. 113^4 should end on $8561 + 8800 = 7361$, and indeed, it is (16304)7361, so this works perfectly.

For the numbers ending on 01, 26, 51 and 76 it is easier: 4×1^3, 26^3, 51^3 and 76^3 will give 04, so the jump of these numbers will be 400 per hundred. Example: $76^4 = (3336) 2176$. For 876 we have a jump of $8 \times 400 = 3200$, and indeed $876^4 = (58886592)\ 5376$, so ending on 5376.

For the numbers ending on 24, 49, 74 and 99 the jump is 4×24^3, 49^3, 74^3 and 99^3; the jump is 9600 per hundred in case of adding or one can choose the other way: subtract 400 per hundred, just what is the most practical for you.

Another thing is also very interesting: numbers which differ by125 or a multiple of it. As a – random – example we take 36. $36^4 = (167)\ 9616$. The jump is $4 \times 36^3 = 100 = $ ....2400. 136^4 ends – according to Newton – on (34210) 2016. Now $61^4 = (1384)\ 5841$. The jump is $4 \times 61^3 \times 100 = $ also 2400. And yes: $161^4 = (67189)\ 8241$, again completely according to Newton.

So to determine the jumps we may conclude that for every number up to 100 the jumps equal the number $4 \times x^3 \times 100$ and that the same jump occurs for $x + 25$, $x + 50$ and $x + 75$.

14.6.2 Three equal final digits

Finding the possibilities is facilitated if one is confident with the squares; up to 1000 is very practical. We start with 001. The origin of a 001 square is either (EH) 001 or (EH) 249. 001 is the first one, the following is the completion to 250, thus 249. ($249^2 = 62\ 001$). The first square ending on $249 = 57$, as $57^2 = 3249$. The completing number will be 193, $193^2 = (37)\ 249$. We now have the quarter 1, 57, 193 and 249. The other candidates can be found by adding 250 to each of the four mentioned and then get 251, 307, 443 and 499. Again adding gives 501, 557, 693, 749 and the last ones are 751, 807, 943 and 999.

The table of the structure of the fourth powers gives us 50 numbers × 16 so we'll have then 800 numbers, out of the 1,000. Where are the missing 200 numbers?

Well, from 0 – 1,000 there are 100 tens, and there are 100 numbers ending on 5. So now we have 800 numbers. These numbers to the fourth power will all result on 0625. So we have to look to more digits for finding differences. More about the fives under the appropriate heading.

14.6.3 Four final digits

Now we'll go to four equal final digits and count up to 10,000. What we find there is similar to what we saw in the three digit affair. We take 0001, which can be the square of (000)1, 1249 or 5001 and 6249. And these numbers in their turn can be the squares of 1, 443, 807 and 1249, four possibilities up to 1250. As from 0 up to 10,000 we go eight steps of 1,250 we may expect that there are 32 candidates which calculated to the fourth power will give 0001, and indeed there are. The other ones are the first ones increased by 1250 or a multiple of it; 1251, 1693, 2947, 2499, 2501, 2943, 3307, 3749, 3751, 4193, 4557, 4999, 5001, 5443, 5807, 6249, 6251, 6693, 7057, 7499, 7501, 9743, 8307, 8749, 8751, 9193, 9557 and 9999.

As we have 125 numbers 0 (mod 16) × 32 we have up to 10,000: 4,000 numbers 0 (mod 16).
As we have 125 numbers 1 (mod 16) × 32 we have up to 10,000: 4,000 numbers 1 (mod 16).
Disregarding the fives we now have counted 8,000 numbers.
The remaining 2,000 come from 1,000 numbers ending in 0 and 1,000 numbers ending in 5.
Now we know the structure of the fourth powers we can start with small numbers and find the "candidates" for the bigger numbers. Simplest example = 1. The candidates are, of course, 1 and its complement to 1250, so 1249. To get 0001 as final figures, we also need a number ending on either 5001 or 0001. From our knowledge of the squares we find resp. 443^2, (19)6249 or 807^2, (65)1249. Now we have 1, 443, 807 and 1249 and can continue by increasing each of these numbers by 1250 and a multiple of it.

We make it more difficult for us and do 1 + (16 × 80) and thus have 1281. Sqrt 1281 gives e.g. 3409, its square being (1162)1281. Sqrt (mod 2) 3409 = 153. Further: 1250 – 153 = 1097; 1250 – 279 = 971. Now we know up to 1250: 153, 279, 971 and 1097. Increasing each of them with 1250 or a multiple of it enables us to find all the all the in question: 1403, 1529, 2221, 2347, 3903, 4029, 4721, 5153, 5279, 5971, 6097, 6403, 6529, 7221, 7347, 7653, 7779, 8471, 8597, 8903, 9029, 9721, 9847.

14.6.4 How to retrieve the basic numbers

Odd numbers

We take at random as a starting point the number 6721. $11^2 = 121$, the hundreds increase by a multiple of 22. After 3 × we have ?6721, $311^2 = 96721$. 311 cannot be a square, but 4689 can. $17^2 = 289$, the hundreds increase by a multiple of 34. We miss 4400 from 289 to 4689; 4400 ÷ 34 Ξ 16. What we have now is 1617, and indeed its square ends on 4689: (261)4689. $1617 - 1250 = 367^2 = (13)4689$. Now we take 367 and its complement 883 to 1250.

There must also be a ??19 number. $19^2 = 361$, $19^4 = 130321$. The jump for $19^4 = 4 \times 19^3 \times 100 = 3600$ per hundred. From 0321 up to 6721 we need an increase of 6400 ÷ 3600 = 24. This means that 2419^4 will result in 6721. This includes that $2500 - 2419 = 81$ will be the number below 1250 which we need, and in the same way we find $1250 - 81 = 1169$. And here we are now, up to 1250 we have 81, 367, 883, 1169. The other numbers up to 10,000 giving 6721 in the fourth power can now easily be found.

Even numbers

Example: 4336. $16^4 = 65536$. The jump for $16^4 = 4 \times 16^3 \times 100 = 8400$ per hundred. From 5536 up to 4336 = 8800. 88 ÷ 84 = possibly 7, 32, 57 or 82. This means that 716, 3216, 5716 and 8216 all result in 4336 when calculated to the 4th power. Up to 1250 we now have already 534 and 716.

Now for the other ones; $12^4 = 20736$. The jump for $12^4 = 4 \times 12^3 \times 100 = (69)1200$ per hundred. From 0736 up to 4336 the difference is 3600 ÷ 1200 Ξ 3, 28, 53 or 78. Now we have 312 as a candidate and his complement to 1250: 938. And of course 2812, 5312, 7812.

Summarised up to 1250 we have, in order of magnitude, 312, 534, 716 and 938. The remaining candidates can now simply be determined by adding 1250 and multiples of it, up to 10,000. What we then get is this series: 312, 534, 716, 938, 1562, 1784, 1966, 2188, 2812, 3034, 3216, 3438, 4062, 4284, 4466, 4688, 5312, 5534, 5716, 5938, 6562, 6784, 6966, 7188, 7812, 8034, 8216, 8438, 9062, 9284, 9446, 9688.

14.6.5 About the fives

Fourth powers ending on 00625.

The following numbers calculated to the fourth power will end on 00625: 5, 35, 45, 85, 115, 155, 165, 195, so 8 per 200. From this point on always + 200 at a time, so 205, 235, 245, 285, 315 et cetera. As we have up to 10,000 50 × 200 there are 400 numbers of which the fourth power ends in 00625.

Fourth powers ending on 40625.

The following numbers calculated to the fourth power will end on 40625: 75, 125, 275,

325, so 4 per 400. From this point on always + 400 each time, so 475, 525, 675, 725, 875, 925, et cetera As we have up to 10,000 25 × 400 there are 100 numbers of which the fourth power is ending on 40625.

Fourth powers ending on 50625.
The following numbers calculated to the fourth power will end on 50625: 15, 55, 65, 95, 105, 135, 145, 185, so 8 per 200. From this point on always + 200, so 215, 255, 265, 295 et cetera. As we have up to 10.000 50 × 200 there are 400 numbers of which the fourth power is ending on 50625.

Fourth powers ending on 90625.
The following numbers calculated to the fourth power will end on 90625: 25, 175, 225, 375, also 4 per 400. From this point on always + 400, so 425, 575, 625, 775, 825, 975 et cetera As we have up to 10,000 25 × 400 there are 100 numbers of which the fourth power is ending 90625.

Altogether we have now 400 + 400 + 100 + 100 = the 1.000 numbers ending on 0625, none of them ending on 10625, 20625, 30625, 60625, 70625 and 80625. Why is that? Each of these numbers mentioned divided by 625 will result in resp. 17, 33, 49, 97, 113 and 129. None of them can be the fourth power of an integer number.

All the fives follow the 1(mod 16) line: (00)625 ÷16 gives a remainder 1 and all the 10,000 are 0 (mod 16) so all the 0625 numbers are 1 (mod 16).

14.6.6 About the tables

The most left column of the next multi-page table gives the four last figures of the fourth power. The other columns present the basic numbers.

0001	1	443	807	1249	1251	1693	2057	2499	2501	2943	3307	3749	3751	4193	4557	4999
0016	2	364	886	1248	1252	1614	2136	2498	2502	2864	3386	3748	3752	4114	4636	4998
0081	3	79	1171	1247	1253	1329	2421	2497	2503	2579	3671	3747	3753	3829	4921	4997
0096	456	492	758	794	1706	1742	2008	2044	2956	2992	3258	3294	4206	4242	4508	4544
0161	391	537	713	859	1641	1787	1963	2109	2891	3037	3213	3359	4141	4287	4463	4609
0176	332	424	826	918	1582	1674	2076	2168	2832	2924	3326	3418	4082	4174	4576	4668
0241	323	589	661	927	1573	1839	1911	2177	3089	3161	3427	4073	4339	4411	4667	4677
0256	4	522	728	1246	1254	1772	1978	2496	2504	3022	3228	3746	3754	4272	4478	4996
0321	19	333	917	1231	1269	1583	2167	2481	2519	2833	3417	3731	3769	4083	4667	4981
0336	284	438	812	966	1534	1688	2062	2216	2784	2938	3312	3466	4034	4188	4562	4716
0401	**101**	257	993	**1149**	**1351**	1507	2243	**2399**	**2601**	2757	3493	**3649**	**3851**	4007	4743	**4899**
0416	298	486	764	952	1548	1736	2014	2202	2798	2986	3264	3452	4048	4236	4514	4702
0481	53	271	979	1197	1303	1521	2229	2447	2553	2771	3479	3697	3803	4021	4729	4947
0496	306	558	692	944	1556	1808	1942	2194	2806	3058	3192	3444	4056	4308	4442	4694
0561	259	263	987	991	1509	1513	2237	2241	2759	2763	3487	3491	4009	4013	4737	4741
0576	218	324	926	1032	1468	1574	2176	2282	2718	2824	3426	3532	3968	4074	4676	4782
0641	261	623	627	989	1511	1873	1877	2239	2761	3123	3127	3489	4011	4373	4377	4739
0656	346	472	778	904	1596	1722	2028	2154	2846	2972	3278	3404	4096	4222	4528	4654
0721	133	169	1081	1117	1383	1419	2331	2367	2633	2669	3581	3617	3883	3919	4831	4867
0736	12	316	934	1238	1262	1566	2184	2488	2512	2816	3434	3738	3762	4066	4684	4988
0801	201	293	957	1049	1451	1543	2207	2299	2701	2793	3457	3549	3951	4043	4707	4799
0816	86	598	652	1164	1336	1848	1902	2414	2586	3098	3152	3664	3836	4348	4402	4914
0881	103	621	629	1147	1353	1871	1879	2397	2603	3121	3129	3647	3853	4371	4379	4897
0896	156	358	892	1094	1406	1608	2142	2344	2656	2858	3392	3594	3906	4108	4642	4844
0961	187	341	909	1063	1437	1591	2159	2313	2687	2841	3409	3563	3937	4091	4659	4813
0976	224	482	768	1026	1474	1732	2018	2276	2724	2982	3268	3526	3974	4232	4518	4776
1041	139	327	923	1111	1389	1577	2173	2361	2639	2827	3423	3611	3889	4077	4673	4861
1056	422	554	696	828	1672	1804	1946	2078	2922	3054	3196	3328	4172	4304	4446	4578
1121	67	319	913	1183	1317	1569	2163	2433	2567	2819	3413	3683	3817	4069	4681	4933
1136	334	462	788	916	1584	1712	2038	2166	2834	2962	3288	3416	4084	4212	4538	4666
1201	301	407	843	949	1551	1657	2093	2199	2801	2907	3343	3449	4051	4157	4593	4699
1216	314	352	898	936	1564	1602	2148	2186	2814	2852	3398	3436	4064	4102	4648	4686
1281	153	279	971	1097	1403	1529	2221	2347	2653	2779	3471	3597	3903	4029	4721	4847
1296	6	158	1092	1244	1256	1408	2342	2494	2506	2658	3592	3744	3756	3908	4842	4994
1361	309	613	637	941	1559	1863	1887	2191	2809	3113	3137	3441	4059	4363	4387	4691
1376	68	124	1126	1182	1318	1374	2376	2432	2568	2624	3626	3682	3818	3874	4876	4932
1441	27	539	711	1223	1277	1789	1961	2473	2527	3039	3211	3723	3777	4289	4461	4973
1456	204	372	878	1046	1454	1622	2128	2296	2704	2872	3378	3546	3954	4122	4628	4796
1521	267	469	781	983	1517	1719	2031	2233	2767	2969	3281	3483	4017	4219	4531	4733
1536	266	338	912	984	1516	1588	2162	2234	2766	2838	3412	3484	4016	4088	4662	4734
1601	143	401	849	1107	1393	1651	2099	2357	2643	2901	3349	3607	3893	4151	4599	4857
1616	52	536	714	1198	1302	1786	1964	2448	2552	3036	3214	3698	3802	4286	4464	4948
1681	71	203	1047	1179	1321	1453	2297	2429	2571	2703	3547	3679	3821	3953	4797	4929
1696	42	144	1106	1208	1292	1394	2356	2458	2542	2644	3606	3708	3792	3894	4856	4958
1761	163	291	959	1087	1413	1541	2209	2337	2663	2791	3459	3587	3913	4041	4709	4837
1776	24	618	632	1226	1274	1868	1882	2476	2524	3118	3132	3726	3774	4368	4382	4976
1841	273	311	939	977	1523	1561	2189	2227	2773	2811	3439	3477	4023	4061	4689	4727
1856	146	322	928	1104	1396	1572	2178	2354	2646	2822	3428	3604	3896	4072	4678	4854
1921	467	619	631	783	1717	1869	1881	2033	2967	3119	3131	3283	4217	4369	4381	4533
1936	112	384	866	1138	1362	1634	2116	2388	2612	2884	3366	3638	3862	4134	4616	4888

The other numbers in these series can easily be found by adding 5000 to the numbers of this page

2001	501	557	693	749	1751	1807	1943	1999	3001	3057	3193	3249	4251	4307	4443	4499
2016	136	248	1002	1114	1386	1498	2252	2364	2636	2748	3502	3614	3886	3998	4752	4864
2081	253	421	829	997	1503	1671	2079	2247	2753	2921	3329	3497	4003	4171	4579	4747
2096	242	294	956	1008	1492	1544	2206	2258	2742	2794	3456	3508	3992	4044	4706	4758
2161	287	359	891	963	1537	1609	2141	2213	2787	2859	3391	3463	4037	4109	4641	4713
2176	76	82	1168	1174	1326	1332	2418	2424	2576	2582	3668	3674	3826	3832	4918	4924
2241	89	573	677	1161	1339	1823	1927	2411	2589	3073	3177	3661	3839	4323	4427	4911
2256	272	496	754	978	1522	1746	2004	2228	2772	2996	3254	3478	4022	4246	4504	4728
2321	481	583	667	769	1731	1833	1917	2019	2981	3083	3167	3269	4231	4333	4417	4519
2336	216	562	688	1034	1466	1812	1938	2284	2716	3062	3188	3534	3966	4312	4438	4784
2401	7	601	649	1243	1257	1851	1899	2493	2507	3101	3149	3743	3757	4351	4399	4993
2416	264	548	702	986	1514	1798	1952	2236	2764	3048	3202	3486	4014	4298	4452	4736
2481	303	479	771	947	1553	1729	2021	2197	2803	2979	3271	3447	4053	4229	4521	4697
2496	442	444	806	808	1692	1694	2056	2058	2942	2944	3306	3308	4192	4194	4556	4558
2561	241	513	737	1009	1491	1763	1987	2259	2741	3013	3237	3509	3991	4263	4487	4759
2576	176	468	782	1074	1426	1718	2032	2324	2676	2968	3282	3574	3926	4218	4532	4824
2641	377	489	761	873	1627	1739	2011	2123	2877	2989	3261	3373	4127	4239	4511	4623
2656	222	404	846	1028	1472	1654	2096	2278	2722	2904	3346	3528	3972	4154	4596	4778
2721	331	383	867	919	1581	1633	2117	2169	2831	2883	3367	3419	4081	4133	4617	4669
2736	238	434	816	1012	1488	1684	2066	2262	2738	2934	3316	3512	3988	4184	4566	4762
2801	543	549	701	707	1793	1799	1951	1957	3043	3049	3201	3207	4293	4299	4451	4457
2816	402	586	664	848	1652	1836	1914	2098	2902	3086	3164	3348	4152	4336	4414	4598
2881	129	353	897	1121	1379	1603	2147	2371	2629	2853	3397	3621	3879	4103	4647	4871
2896	594	608	642	656	1844	1858	1892	1906	3094	3108	3142	3156	4344	4358	4392	4406
2961	63	409	841	1187	1313	1659	2091	2437	2563	2909	3341	3687	3813	4159	4591	4937
2976	232	276	974	1018	1482	1526	2224	2268	2732	2776	3474	3518	3982	4026	4724	4768
3041	77	361	889	1173	1327	1611	2139	2423	2577	2861	3389	3673	3827	4111	4639	4923
3056	54	172	1078	1196	1304	1422	2328	2446	2554	2672	3578	3696	3804	3922	4828	4946
3121	181	183	1067	1069	1431	1433	2317	2319	2681	2683	3567	3569	3931	3933	4817	4819
3136	166	212	1038	1084	1416	1462	2288	2334	2666	2712	3538	3584	3916	3962	4788	4834
3201	157	449	801	1093	1407	1699	2051	2343	2657	2949	3301	3593	3907	4199	4551	4843
3216	102	186	1064	1148	1352	1436	2314	2398	2602	2686	3564	3648	3852	3936	4814	4898
3281	221	403	847	1029	1471	1653	2097	2279	2721	2903	3347	3529	3971	4153	4597	4779
3296	408	506	744	842	1658	1756	1994	2092	2908	3006	3244	3342	4158	4256	4494	4592
3361	191	387	863	1059	1441	1637	2113	2309	2691	2887	3363	3559	3941	4137	4613	4809
3376	318	376	874	932	1568	1626	2124	2182	2818	2876	3374	3432	4068	4126	4624	4682
3441	39	223	2027	1211	1289	1473	3277	2461	2539	2723	4527	3711	3789	3973	4777	4961
3456	122	296	954	1128	1372	1546	2204	2378	2622	2796	3454	3628	3872	4046	4704	4878
3521	17	31	1219	1233	1267	1281	2469	2483	2517	2531	3719	3733	3767	3781	4969	4983
3536	484	588	662	766	1734	1838	1912	2016	2984	3088	3162	3266	4234	4338	4412	4516
3601	349	393	857	901	1593	1643	2107	2151	2843	2893	3357	3401	4093	4143	4607	4657
3616	198	214	1036	1052	1448	1464	2286	2302	2698	2714	3536	3552	3948	3964	4786	4802
3681	453	573	679	797	1703	1823	1929	2047	2953	3073	3179	3297	4203	4323	4427	4547
3696	208	356	894	1042	1458	1606	2144	2292	2708	2856	3394	3542	3958	4106	4644	4792
3761	413	459	791	837	1663	1709	2041	2087	2913	2959	3291	3337	4163	4209	4541	4587
3776	382	476	774	868	1632	1726	2024	2118	2882	2976	3274	3368	4132	4226	4524	4618
3841	439	523	727	811	1689	1773	1977	2061	2939	3023	3227	3311	4189	4273	4477	4561
3856	72	604	646	1178	1322	1854	1896	2428	2572	3104	3146	3678	3822	4354	4396	4928
3921	119	217	1033	1131	1369	1467	2283	2381	2619	2717	3533	3631	3869	3967	4783	4881
3936	116	138	1112	1134	1366	1388	2362	2384	2616	2638	3612	3634	3866	3888	4862	4884

The other numbers in these series can easily be found by adding 5000 to the numbers of this page.

14.6 Structure in the fourth power

4001	249	307	943	1001	1499	1557	2193	2251	2749	2807	3443	3501	3999	4057	4693	4751
4016	498	614	696	752	1748	1864	1946	2002	2998	3114	3196	3252	4248	4364	4386	4502
4081	329	503	747	921	1579	1753	1997	2171	2829	3003	3247	3421	4079	4253	4497	4671
4096	8	206	1044	1242	1258	1456	2294	2492	2508	2706	3544	3742	3758	3956	4794	4992
4161	37	141	1109	1213	1287	1391	2359	2463	2537	2641	3609	3713	3787	3891	4859	4963
4176	168	576	674	1082	1418	1826	1924	2332	2668	3076	3174	3582	3918	4326	4424	4832
4241	411	427	823	839	1661	1677	2073	2089	2911	2927	3323	3339	4161	4177	4573	4589
4256	22	254	996	1228	1272	1504	2246	2478	2522	2754	3496	3728	3772	4004	4746	4978
4321	269	417	833	981	1519	1667	2083	2231	2769	2917	3333	3481	4019	4167	4583	4731
4336	312	534	716	938	1562	1784	1966	2188	2812	3034	3216	3438	4062	4284	4466	4688
4401	149	243	1007	1101	1399	1493	2257	2351	2649	2743	3507	3601	3899	3993	4757	4851
4416	236	452	798	1014	1486	1702	2048	2264	2736	2952	3298	3514	3986	4202	4548	4764
4481	21	553	697	1229	1271	1803	1947	2479	2521	3053	3197	3729	3771	4303	4447	4979
4496	56	192	1058	1194	1306	1442	2308	2444	2556	2692	3558	3694	3806	3942	4808	4944
4561	487	509	741	763	1737	1759	1991	2013	2987	3009	3241	3263	4237	4259	4491	4513
4576	532	574	676	718	1782	1824	1926	1968	3032	3074	3176	3218	4282	4324	4426	4468
4641	11	127	1123	1239	1261	1377	2373	2489	2511	2627	3623	3739	3761	3877	4873	4989
4656	28	96	1154	1222	1278	1346	2404	2472	2528	2596	3654	3722	3778	3846	4904	4972
4721	419	617	633	831	1669	1867	1883	2081	2919	3117	3133	3331	4169	4367	4383	4581
4736	66	488	762	1184	1316	1738	2012	2434	2566	2988	3262	3684	3816	4238	4512	4934
4801	49	457	793	1201	1299	1707	2043	2451	2549	2957	3293	3701	3799	4207	4543	4951
4816	152	164	1086	1098	1402	1414	2336	2348	2652	2664	3586	3598	3902	3914	4836	4848
4881	371	603	647	879	1621	1853	1897	2129	2871	3103	3147	3379	4121	4353	4397	4629
4896	94	392	858	1156	1344	1642	2108	2406	2594	2892	3358	3656	3844	4142	4608	4906
4961	91	313	937	1159	1341	1563	2187	2409	2591	2813	3437	3659	3841	4063	4687	4909
4976	18	474	776	1232	1268	1724	2026	2482	2518	2974	3276	3732	3768	4224	4526	4982
5041	173	389	861	1077	1423	1639	2111	2327	2673	2889	3361	3577	3923	4139	4611	4827
5056	78	446	804	1172	1328	1696	2054	2422	2578	2946	3304	3672	3828	4196	4554	4922
5121	433	569	661	817	1683	1819	1911	2067	2933	3069	3161	3317	4183	4319	4431	4567
5136	38	584	666	1212	1288	1834	1916	2462	2538	3084	3166	3712	3788	4334	4416	4962
5201	51	93	1157	1199	1301	1343	2407	2449	2551	2593	3657	3699	3801	3843	4907	4949
5216	148	564	686	1102	1398	1814	1936	2352	2648	3064	3186	3602	3898	4314	4436	4852
5281	529	597	653	721	1779	1847	1903	1971	3029	3097	3153	3221	4279	4347	4403	4471
5296	244	592	658	1006	1494	1842	1908	2256	2744	3092	3158	3506	3994	4342	4408	4756
5361	137	559	691	1113	1387	1809	1941	2363	2637	3059	3191	3613	3887	4309	4441	4863
5376	374	568	682	876	1624	1818	1932	2126	2874	3068	3182	3376	4124	4318	4432	4626
5441	461	473	777	789	1711	1723	2027	2039	2961	2973	3277	3289	4211	4223	4527	4539
5456	128	454	796	1122	1378	1704	2046	2372	2628	2954	3296	3622	3878	4204	4546	4872
5521	233	531	719	1017	1483	1781	1969	2267	2733	3031	3219	3517	3983	4281	4469	4767
5536	16	412	838	1234	1266	1662	2088	2484	2516	2912	3338	3734	3766	4162	4588	4984
5601	151	607	643	1099	1401	1857	1893	2349	2651	3107	3143	3599	3901	4357	4393	4849
5616	286	448	802	964	1536	1698	2052	2214	2786	2948	3302	3464	4036	4198	4552	4714
5681	179	547	703	1071	1429	1797	1953	2321	2679	3047	3203	3571	3929	4297	4453	4821
5696	394	458	792	856	1644	1708	2042	2106	2894	2958	3292	3356	4144	4208	4542	4606
5761	41	587	663	1209	1291	1837	1913	2459	2541	3087	3163	3709	3791	4337	4413	4959
5776	132	274	976	1118	1382	1524	2226	2368	2632	2774	3476	3618	3882	4024	4726	4868
5841	61	477	773	1189	1311	1727	2023	2439	2561	2977	3273	3689	3811	4227	4523	4939
5856	104	178	1072	1146	1354	1428	2322	2396	2604	2678	3572	3646	3854	3928	4822	4896
5921	33	381	869	1217	1283	1631	2119	2467	2533	2881	3369	3717	3783	4131	4619	4967
5936	388	616	634	682	1638	1866	1884	1932	2888	3116	3134	3182	4138	4366	4384	4612

The other numbers in these series can easily be found by adding 5000 to the numbers of this page.

6001	57	251	999	1193	1307	1501	2249	2443	2557	2751	3499	3693	3807	4001	4749	4943
6016	114	502	748	1136	1364	1752	1998	2386	2614	3002	3248	3636	3864	4252	4498	4886
6081	171	497	753	1079	1421	1747	2003	2329	2671	2997	3253	3579	3921	4247	4503	4829
6096	258	544	706	992	1508	1794	1956	2242	2758	3044	3206	3492	4008	4294	4456	4742
6161	213	609	641	1037	1463	1859	1891	2287	2713	3109	3141	3537	3963	4359	4391	4787
6176	174	418	832	1076	1424	1668	2082	2326	2674	2918	3332	3576	3924	4168	4582	4826
6241	177	339	911	1073	1427	1589	2161	2323	2677	2839	3411	3573	3927	4089	4661	4823
6256	228	246	1004	1022	1478	1496	2254	2272	2728	2746	3504	3522	3978	3996	4754	4772
6321	167	232	1019	1083	1417	1482	2269	2333	2667	2732	3519	3583	3917	3982	4768	4833
6336	34	62	1188	1216	1284	1312	2438	2466	2534	2562	3688	3716	3784	3812	4938	4966
6401	351	493	757	899	1601	1743	2007	2149	2851	2993	3257	3399	4101	4243	4507	4649
6416	202	514	736	1048	1452	1764	1986	2298	2702	3014	3236	3548	3952	4264	4486	4798
6481	447	521	729	803	1697	1771	1979	2053	2947	3021	3229	3303	4197	4271	4479	4553
6496	58	556	694	1192	1308	1806	1944	2442	2558	3056	3194	3692	3808	4306	4444	4942
6561	9	237	1013	1241	1259	1487	2263	2491	2509	2737	3513	3741	3759	3987	4763	4991
6576	74	282	968	1176	1324	1532	2218	2426	2574	2782	3468	3676	3824	4032	4718	4926
6641	123	511	739	1127	1373	1761	1989	2377	2623	3011	3239	3627	3873	4261	4489	4877
6656	278	596	654	972	1528	1846	1904	2222	2778	3096	3154	3472	4028	4346	4404	4722
6721	81	367	883	1169	1331	1617	2133	2419	2581	2867	3383	3669	3831	4117	4633	4919
6736	512	566	664	738	1762	1816	1914	1988	3012	3066	3164	3238	4262	4316	4434	4488
6801	207	451	799	1043	1457	1701	2049	2293	2707	2951	3299	3543	3957	4201	4549	4793
6816	98	336	914	1152	1348	1586	2164	2402	2598	2836	3414	3652	3848	4086	4664	4902
6881	379	397	853	871	1629	1647	2103	2121	2879	2897	3353	3371	4129	4147	4603	4621
6896	142	406	844	1108	1392	1656	2094	2358	2642	2906	3344	3608	3892	4156	4594	4858
6961	563	591	659	687	1813	1841	1909	1937	3063	3091	3159	3187	4313	4341	4409	4437
6976	26	268	982	1224	1276	1518	2232	2474	2526	2768	3482	3724	3776	4018	4732	4974
7041	111	423	827	1139	1361	1673	2077	2389	2611	2923	3327	3639	3861	4173	4577	4889
7056	304	328	922	946	1554	1578	2172	2196	2804	2828	3422	3446	4054	4078	4672	4696
7121	69	567	683	1181	1319	1817	1933	2431	2569	3067	3183	3681	3819	4317	4433	4931
7136	84	288	962	1166	1334	1538	2212	2416	2584	2788	3462	3666	3834	4038	4712	4916
7201	343	551	699	907	1593	1801	1949	2157	2843	3051	3199	3407	4093	4301	4449	4657
7216	64	398	652	1186	1314	1648	1902	2436	2564	2898	3152	3686	3814	4148	4602	4936
7281	29	347	903	1221	1279	1597	2153	2471	2529	2847	3403	3721	3779	4097	4653	4971
7296	256	342	908	994	1506	1592	2158	2244	2756	2842	3408	3494	4006	4092	4658	4744
7361	59	113	1137	1191	1309	1363	2387	2441	2559	2613	3637	3691	3809	3863	4887	4941
7376	126	432	818	1124	1376	1682	2068	2374	2626	2932	3318	3624	3876	4182	4568	4874
7441	289	527	723	961	1539	1777	1973	2211	2789	3027	3223	3461	4039	4277	4473	4711
7456	46	378	872	1204	1296	1628	2122	2454	2546	2878	3372	3704	3796	4128	4622	4954
7521	219	483	767	1031	1469	1733	2017	2281	2719	2983	3267	3531	3969	4233	4517	4781
7536	162	516	734	1088	1412	1766	1984	2338	2662	3016	3234	3588	3912	4266	4484	4838
7601	357	599	651	893	1607	1849	1901	2143	2857	3099	3151	3393	4107	4349	4401	4643
7616	464	552	698	786	1714	1802	1948	2036	2964	3052	3198	3286	4214	4302	4448	4536
7681	297	321	929	953	1547	1571	2179	2203	2797	2821	3429	3453	4047	4071	4679	4703
7696	106	542	708	1144	1356	1792	1958	2394	2606	3042	3208	3644	3856	4292	4458	4894
7761	337	541	709	913	1587	1791	1959	2163	2837	3041	3209	3413	4087	4291	4459	4663
7776	118	226	1024	1132	1368	1476	2274	2382	2618	2726	3524	3632	3868	3976	4774	4882
7841	227	561	689	1023	1477	1811	1939	2273	2727	3061	3189	3523	3977	4311	4439	4773
7856	396	428	822	854	1646	1678	2072	2104	2896	2928	3322	3354	4146	4178	4572	4604
7921	283	369	881	967	1533	1619	2131	2217	2783	2869	3381	3467	4033	4119	4631	4717
7936	134	612	638	1116	1384	1862	1888	2366	2634	3112	3138	3616	3884	4362	4388	4866

The other numbers in these series can easily be found by adding 5000 to the numbers of this page.

8001	193	499	751	1057	1443	1749	2001	2307	2693	2999	3251	3557	3943	4249	4501	4807
8016	252	366	864	998	1502	1616	2114	2248	2752	2866	3364	3498	4002	4116	4634	4748
8081	247	579	671	1003	1497	1829	1921	2253	2747	3079	3171	3503	3997	4329	4421	4753
8096	44	508	742	1206	1294	1758	1992	2456	2544	3008	3242	3706	3794	4258	4492	4956
8161	109	463	787	1141	1359	1713	2037	2391	2609	2963	3287	3641	3859	4213	4537	4891
8176	326	582	668	924	1576	1832	1918	2174	2826	3082	3168	3424	4076	4332	4418	4674
8241	73	161	1089	1177	1323	1411	2339	2427	2573	2661	3589	3677	3823	3911	4839	4927
8256	478	504	746	772	1728	1754	1996	2022	2978	3004	3246	3272	4228	4254	4496	4522
8321	83	519	731	1167	1333	1769	1981	2417	2583	3019	3231	3667	3833	4269	4481	4917
8336	188	466	784	1062	1438	1716	2034	2312	2688	2966	3284	3562	3938	4216	4534	4812
8401	399	507	743	851	1649	1757	1993	2101	2899	3007	3243	3351	4149	4257	4493	4601
8416	14	48	1202	1236	1264	1298	2452	2486	2514	2548	3702	3736	3764	3798	4952	4986
8481	197	229	1021	1053	1447	1479	2271	2303	2697	2729	3521	3553	3947	3979	4771	4803
8496	194	308	942	1056	1444	1558	2192	2306	2694	2808	3442	3556	3944	4058	4692	4806
8561	13	491	759	1237	1263	1741	2009	2487	2513	2991	3259	3737	3763	4241	4509	4987
8576	32	426	824	1218	1282	1676	2074	2468	2532	2926	3324	3718	3782	4176	4574	4968
8641	239	373	877	1011	1489	1623	2127	2261	2739	2873	3377	3511	3989	4123	4627	4761
8656	154	528	722	1096	1404	1778	1972	2346	2654	3028	3222	3596	3904	4278	4472	4846
8721	117	581	669	1133	1367	1831	1919	2383	2617	3081	3169	3633	3867	4331	4419	4883
8736	184	262	988	1066	1434	1512	2238	2316	2684	2762	3488	3566	3934	4012	4738	4816
8801	43	299	951	1207	1293	1549	2201	2457	2543	2799	3451	3707	3793	4049	4701	4957
8816	348	414	836	902	1598	1664	2086	2152	2848	2914	3336	3402	4098	4164	4586	4652
8881	121	147	1103	1129	1371	1397	2353	2379	2621	2647	3603	3629	3871	3897	4853	4879
8896	108	344	906	1142	1358	1594	2156	2392	2608	2844	3406	3642	3858	4094	4656	4892
8961	159	437	813	1091	1409	1687	2063	2341	2659	2937	3313	3591	3909	4187	4563	4841
8976	518	526	724	732	1768	1776	1974	1982	3018	3026	3224	3232	4268	4276	4474	4482
9041	577	611	639	673	1827	1861	1889	1923	3077	3111	3139	3173	4327	4361	4389	4423
9056	196	578	672	1054	1446	1828	1922	2304	2696	3078	3172	3554	3946	4328	4422	4804
9121	317	431	819	933	1567	1681	2069	2183	2817	2931	3319	3433	4067	4181	4569	4683
9136	416	538	712	834	1666	1788	1962	2084	2916	3038	3212	3334	4166	4288	4462	4584
9201	199	593	657	1051	1449	1843	1907	2301	2699	3093	3157	3551	3949	4343	4407	4801
9216	436	602	648	814	1686	1852	1898	2064	2936	3102	3148	3314	4186	4352	4398	4564
9281	97	471	779	1153	1347	1721	2029	2403	2597	2971	3279	3653	3847	4221	4529	4903
9296	92	494	756	1158	1342	1744	2006	2408	2592	2994	3256	3658	3842	4244	4506	4908
9361	363	441	809	887	1613	1691	2059	2137	2863	2941	3309	3387	4113	4191	4559	4637
9376	182	624	626	1068	1432	1874	1876	2318	2682	3124	3126	3568	3932	4374	4376	4818
9441	211	277	973	1039	1461	1527	2223	2289	2711	2777	3473	3539	3961	4027	4723	4789
9456	546	622	628	704	1796	1872	1878	1954	3046	3122	3128	3204	4296	4372	4378	4454
9521	281	517	733	969	1531	1767	1983	2219	2781	3017	3233	3469	4031	4267	4483	4719
9536	88	234	1016	1162	1338	1484	2266	2412	2588	2734	3516	3662	3838	3984	4766	4912
9601	99	107	1143	1151	1349	1357	2393	2401	2599	2607	3643	3651	3849	3857	4893	4901
9616	36	302	948	1214	1286	1552	2198	2464	2536	2802	3448	3714	3786	4052	4698	4964
9681	47	429	821	1214	1297	1679	2071	2464	2547	2929	3321	3714	3797	4179	4571	4953
9696	292	606	644	958	1542	1856	1894	2208	2792	3106	3144	3458	4042	4356	4394	4708
9761	87	209	1041	1163	1337	1459	2291	2413	2587	2709	3541	3663	3837	3959	4791	4913
9776	368	524	726	882	1618	1774	1976	2132	2868	3024	3226	3382	4118	4274	4476	4632
9841	23	189	1061	1227	1273	1439	2311	2477	2523	2689	3561	3727	3773	3939	4811	4977
9856	354	572	678	896	1604	1822	1928	2146	2854	3072	3178	3396	4104	4322	4428	4646
9921	131	533	717	1119	1381	1783	1967	2369	2631	3033	3217	3619	3881	4283	4467	4869
9936	362	366	884	888	1612	1616	2134	2138	2862	2866	3384	3388	4112	4116	4634	4638

The other numbers in these series can easily be found by adding 5000 to the numbers of this page.

14.7 Quantitative powering

We will think about the question: How many figures do we get if we power a given number to a given exponent? This gives a result far from that of the multiplication of the basic number × the exponent.

	Exponent								
B.N.	2	3	4	5	6	7	8	9	10
2	1	1	2	2	2	3	3	3	4
3	1	2	2	3	3	4	4	5	5
4	2	2	3	4	4	5	5	6	7
5	2	3	3	4	5	5	6	7	7
6	2	3	4	4	5	6	7	8	8
7	2	3	4	5	6	6	7	8	9
8	2	3	4	5	6	7	8	9	10
9	2	3	4	5	6	7	8	9	10

B.N. means Basic Number

In a flash you see that the differences are considerable, and it is all very logical; the bigger the number, the more figures with the same exponent. By means of estimation we can get a reasonable approach. E.g. $8^5 = 32768$ which is five figures, so 8^{10} has ten as 8^5 is more than the square root of 1000, which is 31.62. If we round 32768 to 328 and calculate $328^2 = 107584$ then we know that if we multiply this number by 8, we get a bit less than 863, and so conclude that 8^{11} will not have eleven figures, but (only) ten.

Concerning the number 9 we do this: $9^5 = 59049$, which we round to 59050. We square this number and add then get 3486, which we round down to 3485. Now we have an impression about the first figures of 9^{10}. We square this number with the fives trick, so $348 \times 349 = 121452$, the first figures of 9^{20}, rounded to 1214. If we have a four figure number to be multiplied by 9 and we want to have five figures, the multiplier has to be more than 1112, as $1112 \times 9 = 10008$. As 1214 is more than 1112 we know that multiplying 1214 by 9 gives a five figure number and now we know that 9^{21} will have 21 figures and as $1214 \times 81 = 98334$ we know that 9^{22} will also have 21 figures. So after 9^{21} the number 9 jumps.

The number 11 has always intrigued me; it takes a lot of powering before it jumps. My estimation was after the 24th power. After a lot of examination and training one can estimate reasonably closely, but if we want to know exactly – what then? Well, then we arrive at the logarithms, and as this is a part of mathematics in which Dr. B. de Weger plays an important role.

We get a lot of interesting things. E.g.:

$11 = 10^{1.0414}$
$11^2 = 10^{2.0828}$
$11^3 = 10^{3.1242}$
$11^4 = 10^{4.1656}$
$11^5 = 10^{5.2070}$
$11^6 = 10^{6.2484}$
$11^7 = 10^{7.2898}$
$11^8 = 10^{8.3311}$
$11^9 = 10^{9.3725}$
$11^{10} = 10^{10.4139}$
$11^{11} = 10^{11.4553}$
$11^{12} = 10^{12.4967}$
$11^{18} = 10^{18.7451}$
$11^{19} = 10^{19.7865}$
$11^{20} = 10^{20.8279}$
$11^{21} = 10^{21.8692}$
$11^{22} = 10^{22.9106}$
$11^{23} = 10^{23.9520}$
$11^{24} = 10^{24.9934}$
$11^{25} = 10^{26.0348}$

We realise that powering with logarithms means that we multiply the logarithm of a number by the exponent. In this case we only look at the decimals. My estimation was that the number 11 would "jump" after the 24th power, which appears to be correct.

We take log 11 = 1.00414. The question: By which number do we have to multiply this if we want a result > 1? We can see this rather simply: 24 × 0.0414 = 0.9936 and with one time more we are over 1.00. We can also do 1 ÷ 0.0414 = 24.1546.

The next jump will be at 2 × 24.1546 = 48.3092. If we continue with this we see the following.

Jump:

Between 24 and 25
Between 48 and 49
Between 72 and 73
Between 96 and 97
Between 120 and 121
Between 144 and 145
Between 169 and 170, and not between 168 and 169

We now take the number 73, its log is 1.8633. We look only after the decimal part 0.8633. With each multiplication the result is still less than 0.8633 and at a given moment we get a number which is lower than 1 ÷ 0.8633 = 0.115.

We now see:

   1 × 1.86332 = 1.8633 so 73 has 2 figures
   2 × 1.86332 = 3.7266 so 73^2 has 4 figures
   3 × 1.86332 = 5.5900 so 73^3 has 6 figures
   4 × 1.86332 = 7.4533 so 73^4 has 8 figures
   5 × 1.86332 = 9.3166 so 73^5 has 10 figures
   6 × 1.86332 = 11.1799 so 73^6 has 12 figures
   7 × 1.86332 = 13.0433 so 73^7 has 14 figures
   8 × 1.86332 = 14.9066 so 73^8 has 15 figures

We see here a jump back. Because 0.0433, the decimal part of log 73^7 is less than the 0.11 we calculated, we remain in the same ten thousand if we add 0.0433 + 0.8633 = 0.9066 the decimal part of log 73^8. The jump backward takes place after the seventh power.

The turn can be found immediately if we look how far 1.86332 is from the next integer number, 2: 2 / 1.86332 = 0.13668 and we divide 1 by this number: 1 ÷ 0.13668 = 7.3165 so from 7 to 8 there is a turn, a jump. Because 1.86332 is less than 2, it is a jump backward.

Next example, log 31 which is close to the square root of 1000, so the log is close to 1.5.

Log 31 gives us 1.49136 so:

   1 × 1.49136 = 1.4914 so 31   has 2 figures
   2 × 1.49136 = 2.9827 so 31^2 has 3 figures
   3 × 1.49136 = 4.4741 so 31^3 has 5 figures
   4 × 1.49136 = 5.9655 so 31^4 has 6 figures
   5 × 1.49136 = 7.4568 so 31^5 has 8 figures
   6 × 1.49136 = 8.9482 so 31^6 has 9 figures
   7 × 1.49136 = 10.4395 so 31^7 has 11 figures
   8 × 1.49136 = 11.9309 so 31^8 has 12 figures

Etcetera; alternating between 1 more and 2 more, so permanent jumps, but also this pattern is disturbed at a given moment: between 57 and 59 there are two steps of 1.

This is understandable: if the decimal number is lower than 0.5 there comes a moment that there is a disturbance within the same decimal group. Look: 149136 × 57 = 8,500,752; 149136 × 58 = 8,649,588; 149136 × 59 = 8,799,024.

We take another number with a decimal part of about 0.5. 10,000. Log 10,000 = 4. The cube root of 10,000 = 21.5443. Log 21 = 1.32221. Next we calculate 2 × 1.32221 =

2.64444; 3 × 1.32221 = 3.96666; 4 × 1.32221 = 5.28888, so after 3× it jumps.

In a calculation tournament there was this question: 97^{97}; has this number 193 or 194 figures? Well, log 97 = 1.98677. Then 2 − 1.98677 = 0.01323, and 1 ÷ 0.01323 = 75.5857, so the turning point is 75. Then: 97^{97} has "only" 193 figures. The next turning point is 151 because 2 ÷ 0.01323 = 151.1715.

15 Integer fourth roots with mod 101

In the case of a fourth root and going from 1 up to 1,000 there are 32 different possibilities. E.g. for 0096 there are 456, 492, 758, 794, 1706, 1742, 2008, 2044, 2956, 2992, 3258, 3294, 4508, 4544, 5456, 5492, 5758, 5794, 6706, 6742, 7008, 7044, 7956, 7992, 8258, 8294, 9206, 9242, 9508, 9544.

Which number was to be used as a modulo? 11? No, the number is too small. 37? In fact this is too small too. I came to 101. Before now I hardly used this number, in fact this is a very good expedient after making myself confident with it. Up to 99 we have a permanent repetition of two 2 digit numbers, so if we divide a number from right to left in groups of four digits, this will work.

The table works as follows: in the column BN^4 you see the mod 101 of the basic number, calculated to the fourth power. In the four columns right of it, you see the mod 101 of the possible basic numbers.

E.g. in the left column is 5 (mod 101) of which basic number? One of them is 34, because $34^2 = 1156 = 45$ (mod 101) and $45^2 = 2025 = 5$ (mod 101). If you do the complement you immediately know that if 34 is one possibility, then 67 is another one.

$37^2 = 1369 = 56$ (mod 101), the complement of 45 (mod 101), $56^2 = 3136 = 5$ (mod 101) and again $5^2 = 25$ (mod 101). The other numbers work in the same way. We have here a table of twenty five rules and the table is our expedient for extracting integer fourth roots.

What is the mod 101 of 2231? Well, 22 22 = 0 (mod 101) and the mod 101 of the

BN^4 mod 101	BN mod 101			
1	1	10	91	100
5	34	37	64	67
16	2	20	81	99
19	5	50	51	96
24	15	49	52	86
25	45	46	55	46
31	12	19	82	89
36	14	39	62	87
37	18	22	79	83
52	26	43	58	75
54	4	40	61	97
56	8	21	80	93
58	25	48	53	76
68	35	47	54	66
71	23	28	73	78
78	7	31	70	94
79	13	29	72	88
80	27	33	68	74
81	3	30	71	98
84	6	41	60	95
87	36	44	57	65
88	16	42	59	85
92	24	38	63	77
95	17	32	69	84
97	9	11	90	92

number is + 9. 44 08 = 44 44 − 36 so the mod 101 of this number is, just what you prefer, -36 or +65.

E.g. 2231440831420641 gets 2231 | 4408 | 3142 | 0641. 2231 = + 9 (mod 101), 4408 is either -36 or +65 (mod 101), 3142 is +11 (mod 101) and 0641 is +35 (mod 101), so the whole number is +19 (mod 101).

101 is the sum of two different squares, therefore it may be that in the fourth power there are four numbers with the same modulus, just as there are for example with 13. 2^4, 3^4, 10^4 and 11^4 are each 3 (mod 13), something like that should also occur with 101. And indeed it does: 1^4, 10^4, 91^4 and 100^4 are all 1 (mod 101). There is a simple 10 × relation between these numbers: 1 × 10 = 10, 10 × 10 = 100 and 100 × 10 = 91 (mod 101). This lightens the labour considerably.

15.1 How many figures are to be calculated?

An interesting question is: How many figures of the answer are we have to calculate before we can obtain the complete answer by reasoning?

Therefore randomly three numbers were chosen of which the calculation will be elaborated.

A word of advice: Take care that in the fourth roots the number of figures minus two are calculated and then find the two remaining figures by reasoning.

15.2 At work!!

To start with, the fourth root of

$$2231\ 4408\ 3142\ 0641$$

So to the integer fourth root of 2231 | 4408 | 3142 | 0641. 2231 = + 9 (mod 101), 4408 = + 65; 3142 = + 11; 0641 = + 35; the sum is 120 = 19 (mod 101) of the whole number.

On the table we find for 19 (mod 101) the following possibilities for the basic number: 5, 50, 51 and 96. So the result should be one of these possibilities.

Sqrt 2231 4408 = ± 472390. Sqrt 472390 = ± 6873.

Now the possibilities. Because we work here in the hundreds atmosphere the possibilities for ending on 41 are: 11, 39, 61, 89, 23, 27, 73, 77. Now we'll combine with 68 and get 6811 = -57 (mod 101), 6839 = -29 (mod 101), 6861 = -7 (mod 101), 6889 = 21 (mod 101), 6823 = -45 (mod 101) , 6827 = -41 (mod 101), 6873 = 5 (mod

101) and finally 6877 = 9 (mod 101). The only one which accords to the specifications is 6873, and this is the correct result.

What we'll do is to use bigger numbers and examine if there is a chance that mod 101 makes it necessary to combine with another modulo. We'll see

We now take

$$102\ 1834\ 7776\ 3236\ 6096$$

which is a 19 digit number, and so has a 5 digit answer. We subdivide this as follows:

102 | 1834 | 7776 | 3236 | 6096. The first digits are easy to see: $31^4 = 923521 + 12$ zeros, $32^4 = 1048576 + 12$ zeros, so the answer begins with 31. Sqrt 1,021,834 = ± 1010.85. Sqrt 1010.85 is a bit lower than 318, so 317+.

The mod 101 of the whole number is $1 + 16 + -1 + 4 + 36 = 56$. In the table we find as possibilities: 8, 21, 80 and 93.

For completeness we'll examine all the 31,000 candidates, although we know that from 31,500 is enough. So 31,044 = 37 (mod 101), 31,206 = 98 (mod 101), 31,294 = 85, 31,456 = 45, 31,544 = 32, 31,706 = 93, 31,794 = 80, 31,956 = 40, 31,008 = 1, 31,242 = 33, 31,256 = 49, 31,492 = 81, 31,508 = 97, 31,742 = 28, 31,758 = 44, 31,794 = 80, 31,992 = 76, all modulo 101.

The only number which fits is 31,794, we may now conclude that this number which is 80 (mod 101) is the correct answer.

Obviously my method works; I do one more work-out of the following 19 digit number.

$$120\ 0651\ 4159\ 2369\ 9216.$$

Sqrt 1,200,651 = ± 1095.75. Sqrt 1095.75 = ± 33.100. The possibilities: are to be found in the table.

Basic number	Mod 101	Basic number	Mod 101
331 02	-26	331 14	-14
331 48	20	331 36	8
331 52	24	331 64	36
331 98	70	331 86	58

Starting with my knowledge of the squares I do the 4th powers and know that in the 100 group only the numbers 102, 148 and 186 will result in the three last digits being 216.

To find the mod 101 of a five figure number is rather easy, if you split the number as follows: 3 31 02. The three represents the ten-thousands, so positive; the 31 represents the hundreds, you take either -31 or +70, the 02 represents the units, also positive. We get either 3 + 70 + 02 = + 75 or its opposite – 26.

The mod 101 of the whole number is 19 + 45 + 18 + 46 + 25 = 52. The table gives, for 52 mod 101, the following possibilities: 26, 43, 58, 75. As 33102 = -26 this is the one we were looking for, so answer of the question is 33 102.

Encouraged by the successes we'll make some steps and want to do integer fourth roots of numbers with basic numbers of 8 digits, so from 29 up to 32 figures.

15.3 Extracting integer fourth roots from 29 up to 32 digit numbers

Question number	Answer from up to
29 digits	10 00 00 00 up to 17 78 27 94
30 digits	17 78 27 94 up to 31 62 27 77
31 digits	31 62 27 77 up to 56 23 41 33
32 digits	56 23 41 33 up to 99 99 99 99

How many digits??

A very interesting question is: How many digits of the answer do we need to calculate before we go on with reasoning to find the complete answer? To find this I took three randomly chosen numbers and worked it all out as follows. The fourth root of:

$$11\ 1113\ 0200\ 3376\ 0452\ 9085\ 6569\ 0081$$

The question number has 30 digits, the answer will be 18 xx xx xx. Sqrt 11 1113 0200 = 33,333.62; sqrt 33,333.62 = 182.574. Answer 18 25 74 xx. Now reasoning: Mod 101; 18 and 74 are negative: 18 00 00 00, 74 00. – 18 – 74 = -92 or (easier to work with) +9. 25 = +; so 18 25 74 00 = +34 (mod 101).

The mod 101 = 11 + 2 – 2 + 43 + 48 – 5 + 4 + 81, overall + 81 (mod 101)

As the question number is 11 + 2 – 2 + 43 + 48 – 5 + 4 + 81 = 81 (mod 101) the answer is one of the following possibilities: 3, 30, 71 or 98 (mod 101). We look to see if only the first four digits of the answer will satisfy for finding the complete answer. 18 25 = +7 (mod 101). For 0081 as the last four digits we make this table for the basic numbers and their mod 101:

0003	+3	2503	-22	5003	-47	7503	-72
0079	+79	2579	+54	5079	+29	7579	+4
1171	+60	3671	+35	6171	+10	8671	-15
1247	+35	3747	+10	6247	-15	8747	-40
1253	+41	3753	+16	6253	-9	8753	-34
1329	+16	3829	-9	6329	-34	8829	-59
2421	-3	4921	-28	7421	-53	9921	-78
2497	+73	4997	48	7497	23	9997	-2

Is there or are there in this table numbers to be combined with the + 7 of 1825 which result in one of the four possible moduli? Yes: 7497 = + 23, + 7 = 30 (mod 101) and in this case the correct answer.

Now the fourth root of

9755 1123 4995 0157 9255 9730 6157 7041

The mod 101 of the question number is − 42 + 12 + 46 + 56 − 37 − 67 − 4 − 29 = 36 (mod 101), so the basic number can be 14, 49, 62 or 87 (mod 101).

We can restrict ourselves to firstly calculate the square root of 9755 1123 5000, a minor rounding, which does not affect our accuracy. The answer is ± 987,678. Our second step is the square root of 987,678 which is ± 993.820. We now have in the meantime six figures of our answer! The complete answer is 99 38 20 xx. In the table hereunder we see that in the 20 range there is only one number and therefore the answer is 99 38 20 77.

0111	+10	2611	-15	5011	-39	7611	-65
0423	+19	2923	-6	5423	-31	7923	-56
0827	+19	3327	-6	5827	-31	8327	-56
1139	+28	3639	+3	6139	-22	8639	-47
1361	+48	3861	+23	6361	-2	8861	-27
1673	+57	4173	+32	6673	+7	9173	-18
2077	+57	4577	+32	7077	+7	9577	-18
2389	+66	4889	+41	7389	+16	9889	+9

Also in this table we see 8 "pairs" of numbers with the same moduli, besides they lie close to each other, as e.g. 0423 and 0827; and 1673 and 2077. Therefore it is necessary to calculate the first six digits of the answer ourselves and then, with the help of the table, find the last ones by reasoning.

Now we do sqrt of 97551123 = 9876.7972; of which the sqrt is 99.3820. A look at the table shows us that 99 38 20 77 is the correct answer.

Finally the fourth root of

$$325\ 4900\ 2295\ 0737\ 6550\ 5519\ 9396\ 0401$$

The question number has 31 figures, so the answer has eight figures. The question number is +22 − 49 + 73 + 30 − 15 − 36 + 3 − 3 = 25 (mod 101) so the question number will be one of the following possibilities; 45, 46, 55 or 56 (mod 101).

Sqrt 32549002295 = 180413; sqrt 180413 = 424,751.

42 47 51 00 = − 46 = + 55 (mod 101). Hereunder you'll find the possibilities:

0101	0	2601	+76	5101	+51	7601	+26
0257	+55	2757	−30	5257	+5	7757	−20
0993	+84	3493	+59	5993	+34	8493	+9
1149	+38	3649	+13	6149	−12	8649	−37
1351	+38	3851	+13	6351	−12	8851	−37
1507	−8	4007	−33	6507	−58	9007	−83
2243	+21	4743	−4	7243	−29	9743	−54
2399	+76	4899	+51	7399	+26	9899	+1

What we see here is very tricky: 1149 and 1351 close to each other and the same mod 101. Furthermore we see 2399 and 2601, both + 76 (mod 101), 4899 and 5101, both + 51 (mod 101), 6149 and 6351, both -12 (mod 101), 7399 and 7601, both + 26 (mod 101).

This is too critical, so calculating six digits of the answer here is a necessity. As we have already, 42 47 51 01 gives the correct answer.

15.4 Questions; integer fourth roots

$$4\ 8886\ 9757\ 3308\ 3298\ 6518\ 2006\ 2736$$
$$23\ 9855\ 8005\ 3223\ 8364\ 5497\ 6836\ 4961$$
$$875\ 7272\ 6302\ 5125\ 5660\ 0581\ 4412\ 3761$$
$$3214\ 8573\ 1505\ 3414\ 0641\ 6039\ 4959\ 5441$$

15.5 The answers

$$4\ 8886\ 9757\ 3308\ 3298\ 6518\ 2006\ 2736.$$

Square root of 4 8886 9757 = ± 22110. Square root of 22110 = ± 148.69. We now have the fourth root of the first nine figures of the question number and already five figures of the complete answer. We go to the table of the last four figures of the number 2736, select the group in the nine thousand and have 9184, 9566 and 9762.

The mod 101 of the question number is 52 and according to the mod 101^4 table these are the possibilities; 26, 43, 58 and 75.

The first four figures of the answer are 1486 = 72 (mod 101).

From 72 to

   26 we need 55 or -46
   43 we need 72 or -29
   58 we need 87 or -14
   75 we need 03 or -26

9184 = -7, so it drops out. 9566 = -29 + 72 = 43, this is the right choice and our final answer is **1486 9566**.

Next question number:

$$23\ 9855\ 8005\ 3223\ 8364\ 5497\ 6836\ 4961$$

The question number has thirty figures, so the answer has eight. Square root of 23 9855 8000 = ± 48975 and sqrt of 48975 = ± 221.30. These are the first five figures of the answer. The final answer has this shape: 2213 0x xx. This part of the answer is -9 or 92 (mod 101).

Now we go to the table of the last four figures and select 4961 and take out the possibilities the 0000 group: 0091, 0313 and 0937.

The mod 101 of the question number is 1 and according to the mod 101^4 table these are the possibilities: 1, 10, 91 and 100.

The first four figures of the answer are 2213 = -9 (mod 101) or 92.

From -9 or 92 to

   01 we need 10 or -91
   10 we need 19 or -82
   91 we need 100 or -10
   100 we need 08 or -93.

We see that 0313 with +10 (mod 101) is the only number which fits and so the final answer is **2213 0313**.

Next question number:

$$875\ 7272\ 6302\ 5125\ 5660\ 0581\ 4412\ 3761$$

The question number has 31 figures, so the answer has 8.

Square root of 8757272, seven figures will do, 2959.26 and the square root of this number gives 54.399. With this step we have already 5439, the first four figures of the answer and the fifth one is 9. Answer so far 5439 9xxx.

Now we go to the table of the last four figures and select 3761 and take out the 32 possibilities the 9000 group: 9163, 9209, 9541 and 9587.

The mod 101 of the question number is 52 and according to the mod 101^4 table these are the possibilities: 26, 43, 58 and 75.

The first four figures of the answer are 5439 = -15 or 86 (mod 101).

From -15 or 86 to

26 we need 41
43 we need 58
58 we need 73
75 we need 90

Our candidates: 9163 = -28 or 73; 9209 = -83 or 18; 9541 = -54 or 47; 9587 = -8 or 93. The only number which fits is 9163 and therefore our final answer is **5439 9163**.

And the last one:

$$3214\ 8573\ 1505\ 3414\ 0641\ 6039\ 4959\ 5441$$

Square root of 32148573 = 5669.71 and the square root of 566971 = 75.299. Answer so far 75.29, our first four figures and the group of the last four figures starts with 9.

Now we go to the table of the last four figures and select 5441 and take out the 32 possibilities the 9000 group; 9211, 9223, 9527 and 9539.

The mod 101 of the question number is 52 and according to the mod 101^4 table these are the possibilities: 26, 43, 58 and 75.

The first four figures of the answer are 7529 = -46 (mod 101) or 55.

From -46 or 55 to

26 we need 72
43 we need 89
58 we need 03
75 we need 20

Our candidates: 9211 = -81, or 20; 9223 = -69 or 32; 9527 = -68 or 33; 9539 = -56 or 45. The only number which fits is 9211 = -81 or + 20 and therefore our final answer is **7529 9211**.

15.6 The fives in the fourth power

As the fives behave different from the other numbers you will not be amazed that we, when powering, have to take a bigger number of digits than usual. With the other numbers we can see a lot of things when considering the last four figures in the fourth, for the fives we are obliged to take one more step: the last six digits. If we take only four final figures we see only 0625, but there is more.

In this chapter we will look for a possibility to find the fourth root of a big number "in the fives" by means of reasoning based on the recognition of the final figures. For the fives four final figures will not do, we need six of them.

To obtain a survey about this all, all the numbers ending on 5 up to 400 were calculated to the fourth power. We give the last six figures of the results. The table shows us these.

5	000625	35	500625	15	050625	55	150625
85	200625	45	100625	65	850625	105	550625
155	200625	115	900625	95	450625	135	150625
165	200625	195	900625	145	050625	185	350625
235	800625	205	100625	255	250625	215	750625
245	000625	285	500625	305	650625	265	550625
315	600625	355	300625	335	450625	295	350625
395	800625	365	900625	385	650625	345	950625
405	200625	435	100625	415	450625	455	350625
485	800625	445	900625	465	250625		
				495	250625		
125	140625	25	390625	75	640625	175	890625
275	140625	375	390625	325	640625	225	890625
		425	390625	475	640625		

We start with the first part: the column starting with 5, and then 85, 155, 165 et cetera

Typical for the final figures is they end on 00625 and the hundred thousands are even. Look at the "pair" 155 and 165 and the "pair" 235 and 245. Here we find there is a multiple of 80 and a plus and minus 5. However: there is one exception: if 80 plus or minus five leads to a number ..25 or ..75 as we later on will see, it is different. Therefore we miss in this part the numbers 75 and 325.

The part which starts with 35, 45, 115 and so on. Typical for the final figures is that they end on 00625 and that the hundred thousands are odd. The "pairs" here are 35 and 45, then 115. 125 drops out as it ends on 25, later on we have the "pair" 195 and 205 et cetera. Here we find a multiple of 40 plus or minus five, and later on increased by 80.

The group 15, 65, 95 and so on. Here the pairs are 15 / 65 and 95 / 145. The numbers end on 50625, all with an even hundred thousand. The leading number here is 80, with a plus and minus of 15. We do not find here 175 and 225, as they end on 25 or 75.

Then the group 55, 105, 135 et cetera. There the hundred-thousands all are odd. Again the leading number is 80, with a plus and minus of 25.

125 and 275: their fourth powers end on 140625, the hundred thousand is odd. The leading number is an even hundred, here it is 200, then plus or minus 75. The "jump" for the hundreds here is 400; the numbers 525 and 675 calculated to the fourth power end equally on 140625.

The group 25, 375 and 425. Their fourth powers end on 390625, so an odd hundred thousand. The leading number is 400, and plus or minus 25. If we continue with 400 more and then see what happens with 775 and 825: calculated to the fourth power they also end on 390625.

Now the group with 75, 325 and 475. The last figures are 640625, an even hundred thousand. Here we see again a line: they are a multiple of 400, then plus or minus 75. The hundred thousands are even. For testing I calculated 725 and 875 to the fourth power, and according to the expectations these results also end on 640625.

The last group: 175 / 225. Their fourth powers end on 890625, so an even hundred thousand. The central number here 200 with plus and minus 25. For all security the fourth powers of 575, 625 and 975 were calculated, without any exception the last six figures were 890625.

16 Structures in the fifth power

The work out for the formula $(a+b)^5$ is as follows: $a^5 + 5a^4b + 10a^3b^2 + 10a^2b^3 + 5ab^4 + b^5$.

Table 1

BN	BN^5	1 LF	2 LF
1	1	1	01
2	3	2	32
3	24	3	43
4	102	4	24
5	312	5	25
6	777	6	76
7	1680	7	07
8	3276	8	68
9	5904	9	49

We will work out the changes in the numbers in the tens, hundreds and thousands. The way of working will be that the units are always 'b' and that the tens, hundreds and thousands will be 'a'.
Generally it is known that in the fifth power the last figures of a number remain unchanged: 1^5 ends on 1, 2^5 ends on 2 et cetera.

LF means last figures

However, if we take more than one last figure the differences will come.
We'll see that when we add tens to these individual figures the last figures change, dependent of the basic figures.

Table 2

BN	BN^5	BN+10^5	BN+20^5	BN+30^5	BN+40^5
1	01	51	01	51	01
2	32	32	32	32	32
3	43	93	43	93	43
4	24	24	24	24	24
5	25	75	25	75	25
6	76	76	76	76	76
7	07	57	07	57	07
8	68	68	68	68	68
9	49	99	49	99	49

What we see here is the following: The odd basic numbers change by 50 when increasing by 10, the even numbers retain their tens.

The explanation for this phenomenon is this: B is the last figure of the basic number and determines the last two figures. It is $5ab^4$ which determines the tens in the two last figures. If a is an odd ten then we have 5 × 10 × odd ten which is always 50. With 3 e.g. we see when calculating to the fifth power 243, 293, 343, 393, 443 et cetera

In case of an even basic number, with $5ab^4$ we have $5 \times 10 \times 2^4$ which is 800. So if we work with an even basic number we see that the hundreds increase with an even hundred. We'll examine how many last figures we can calculate by head, for sake of interest.

16.1 Numbers ending on 1

BN	BN⁵,4 LF
1	1
11	1051
21	4101
31	9151
41	6201
51	5251
61	6301
71	9351
81	4401
91	1451
101	0501
111	1551

Table 3

What we see here is the increase of the tens, with 50, on the base of $5ab^4$ where b = 1 and a = 10, which is 50. Herewith the increase of the tens is explained, and we can find the last three figures of the numbers.

The thousands, and so the last four figures are determined by $10a^2b^3$. As a = 10, a^2 = 100 and by consequence $10a^2$ = 1000. This means that the thousands are determined by b^2. For a = 20 we get $20a^2b^3$ = 4000, in the same way $30a^2b^3$ = 9000, and if you follow the right column you'll see that the thousands increase just as the squares of the tens. E.g. 71^5: 001 + 7 × 50 = 351, 7^2 ends on 9, so 71^5 ends on 9351.

The "cadence" in the last four figures is: 1050; 3050; 5050; 7050; 9050 and again 1050.

If you want to know which ones are the last four figures of 391^5 simply start with 91^5 and add 5 × 300 = 1500, and look; indeed the last four digits of 391^5 are 2951.

16.2 Numbers ending on 2

Table 4

			3	2
b^5				
$5ab^4$		8	0	0
$10a^2b^3$	8	0	0	0
	8	8	3	2

We work again with $5ab^4$, where b = 2 and a = 10.
For 12 = $5ab^4$ = 5 × 10 × 16 = 800. As the tens increase by ten we'll see that the hundreds increase by 8, so following the third figure from the right there is an increase of 800.

For finding the thousands we use the part: $10a^2b^3$. As b = 2 we have 8 × 10 = 80 to multiply by a^2. So for a = 10 we get 10^2 × 80 = 8000, and now we have the addition according to table 4.

Table 5

			3	2
b^5				
$5ab^4$		1	6	0	0
$10a^2b^3$	2	0	0	0
	3	6	3	2

For the number 22 we use again the part: $10a^2b^3$, which means that $10a^2$ = 4000 × b^3 = 8 results in 32000. Herewith the way to find to four last figures of x2 numbers to the fifth power is explained.

Table 6

BN	BN^5,4 LF
2	32
12	8832
22	3632
32	4432
42	1232
52	4032
62	2832
72	7632
82	8432
92	5232
102	8032
112	6832
122	1632

The work out for the last four figures of 22^5 goes the same way where b remains 2 and a will be 20, so $5ab^4$ = 5 × 20 × 16 = 1600. The cadence is 8800; 4800; 800; 6800; 2800; 8800 et cetera

16.3 Numbers ending on 3

BN	BN⁵, 4 LF	BN	BN⁵,4 LF
3	243	63	6543
13	1293	73	1593
23	6343	83	0643
33	5393	93	3693
43	8443	103	0743
53	5493	113	1793

Table 7

For the threes as usual b = 3 and the tens are represented by the a. $13 = 5ab^4 = 5 \times 10 \times 81 = 4050$. This explains the increase of the tens going with (40) 50, per ten. So also the thousands increase.

The other part of the thousands comes from $10a^2b^3$. For 13 $10a^2 = 1000 \times b^3 = 27000$.

Table 8 a = 10, b = 3

b^5			2	4	3
$5ab^4$		4	0	5	0
$10a^2b^3$	(2)	7	0	0	0
		1	2	9	3

Table 9 a = 20, b = 3

b5			2	4	3
5ab4		8	1	0	0
$10a^2b^3$	(0)	8	0	0	0
		6	3	4	3

From 243 the cadence is + 1050; 5050; 9050; 3050; 7050; 1050

16.4 Numbers ending on 4

BN	BN⁵, 4 LF	BN	BN⁵,4 LF
4	1024	64	1824
14	7824	74	6624
24	2624	84	9424
34	5424	94	0224
44	6224	104	9024
54	5024	114	5824

Table 10

As $5b^4 = 5 \times 256 = 1280 \times 10 = 12800$ we see an increase of the hundreds with (mod 2) 800. $10b^3 = 640, \times 10 = 4000$, therefore 7824.

The cadence is 6800; 4800; 2800; 800; 8800; 6800 et cetera.

As $5b^4 = 5 \times 256 = 1280 \times 20 = 25600$, herewith the hundreds 6 are explained. $10b^3 = 6\ 40, \times 20^2 = (25)6000$, so that the last four figures will be 2624, see table 12.

Table 11 a = 10, b = 4

b^5		**1**	**0**	2	4
$5ab^4$	(1)	**2**	**8**	0	0
$10a^2b^3$	(6)	**4**	**0**	0	0
		7	**8**	2	4

Table 12 a = 20, b = 4

b^5		**1**	**0**	2	4
$5ab^4$	(2)	**5**	**6**	0	0
$10a^2b^3$	(5)	**6**	**0**	0	0
		2	**6**	2	4

16.5 Numbers ending on 5

This is very evident: as every number ending on 5 is a multitude of five, we will retrieve this when we are powering. Also, if we add two numbers which sum is 80 we see that the sum is 50,000. Take e.g. the fifth power of 5 + 75: their sum is 50,000, in the same way the other "pairs".

We may presume that if we add two numbers which sum is 160, the sum of the last five figures will be 100,000.

Table 13

BN	BN^5, 5 LF	BN	BN5, 5 LF
5	03125	65	90625
15	59375	75	46875
25	65625	85	53125
35	21875	95	09375
45	28125	105	15625
55	84375	115	71875

Let's see; 55 and 105: 55^5 = (5032) 84375 + 105^5 = (127628) 15625 = 100,000. We can do the same with 35^5 and 125^5: 21875 + 78125 = 100,000.

16.6 Numbers ending on 6

BN	BN^5, 4 LF	BN	BN^5, 4 LF
6	7776	66	2576
16	8576	76	5376
26	1376	86	0176
36	6176	96	6976
46	2976	106	5776
56	1776	116	6576

Table14

As $5b^4$ = 5 × 1296 = 6480 × 20 = 129600, we see that there is an increase of 4800 per ten. $10a^2b^3$ = 4000 × b^3, 216 , so the thousands increase in total by 3600, the result is the answer 1376 for the last four figures.

Table 15 a = 10, b = 6

b^5			7	7	7	6
5ab^4		(6)	4	8	0	0
10a^{2}b^3	(2)	(1)	6	0	0	0
			8	5	7	6

Table 16 a = 20, b= 6

b^5		7	7	7	6
5ab^4	(2)	9	6	0	0
10a^{2}b^3	(6)	4	0	0	0
		1	3	7	6

The cadence is 800; 2800; 4800; 6800; 8800; 0800 et cetera.

16.7 Numbers ending on 7

BN	BN^5,4 LF	BN	BN^5,4 LF
7	6807	67	5107
17	9857	77	4157
27	8907	87	9207
37	3957	97	0257
47	5007	107	7307
57	2057	117	0357

Table 17 a = 10, b = 7

As $7^4 = 2401$ with $5ab^4$ the tens increase by (0)50 per ten.
a = 10, b = 7, so $5ab^4$ = 120050, the tens increase by 050 per 10.
Here $10b^2 = 1000$, × 7^3 = (34)3000.
The cadence is 3050; 9050; 5050; 1050; 7050; 3050. The increase of the thousands comes from the change of a^2, each time × b^3, 343.

Table 18 a = 10, b = 7

b^5		1	6	8	0	7
$5ab^4$	1	2	0	0	5	0
$10a^2b^3$	3	4	3	0	0	0
			9	8	5	7

Table 19 a = 20, b = 7

b^5		1	6	8	0	7
$5ab^4$	2	4	0	1	0	0
$10a^2b^3$	3	7	2	0	0	0
	(6)	(2)	8	9	0	7

16.8 Numbers ending on 8

BN	BN^5,4 LF	BN	BN^5, 4 LF
8	2768	68	3568
18	9568	78	4368
28	0368	88	9168
38	5168	98	7968
48	3968	108	0768
58	6768	118	7568

Table 20

The cadence is 6800; 0800; 4800; 8800; 2800; 6800 et cetera.

As $b^4 = 4096 × 5 × 10 = 204800$ we recognise the increase of the hundreds by 8 per ten.

Table 21 a = 10, b = 8

b^5		3	2	7	6	8
$5ab^4$	2	0	4	8	0	0
$10a^2b^3$	5	1	2	0	0	0
		(3)	9	5	6	8

Table 22 a = 20, b= 8

b^5		3	2	7	6	8
$5ab^4$		4	0	9	6	0	0
$10a^2b^3$	2	0	4	8	0	0
			0	3	6	8

16.9 Numbers ending on 9

BN	BN^5,4 LF	BN	BN^5,4 LF
9	9049	69	1349
19	6099	79	6399
29	1149	89	9499
39	4199	99	0499
49	5249	109	9549
59	4299	119	6599

Table 23 a = 10, b = 9

As $5b^4 = (32)8050$ the increase of the tens with 050 is clear. In the thousands because of the 8 we have already + 8050 per ten.

Table 24 a = 10, b = 9

b^5		5	9	0	4	9
$5ab^4$	3	2	8	0	5	0
$10a^2b^3$	7	2	9	0	0	0
	(1)	(1)	6	0	9	9

Table 25 a = 20, b = 9

b^5		5	9	0	4	9
$5ab^4$	6	5	6	1	0	0
$10a^2b^3$	9	1	6	0	0	0
			1	1	4	9

The cadence – table 23 – is 7050; 5050; 3050; 1050; 7050 et cetera.

Here $10a^2 = 4000$, to be multiplied by 9^3 ($10a^2b^3$), meaning in combination with $5ab^4$ a total increase of the thousands from 6 to 1(149).

Table 26

BN	BN^5	Cadence
9	59 049	
29	20 511 149	2100
49	282 475 249	4100
69	1564 031 349	6100
89	5584 059 449	8100
109	15386 239 549	0100

16.10 The end of endless…

As there is no end to the numbers, neither there is an end to powering. Except the number 1 which remains always the same, we can say about the other numbers:

- all the even numbers end on 9376, depending on how long we continue, more on this later on
- all the odd numbers end on 0001, depending on how long we continue, more on

this later on

- all the numbers ending on 5 will end on 0625 of 50625, if calculated to the fourth power
- all the numbers ending on 25 will end on 390625 or 890625, if calculated to the fourth power.

There are numbers with which the last figures can be found faster, here are some examples:

Table 27

BN	10	20	100	500
11, 21, 31	4601, 8201, 0801	9201, 6401, 1601	6001, 2001, 8001	0001, 0001, 0001
03, 13, 23	9049, 1849, 3649	4401, 8801, 5201	2001, 4001, 6001	0001, 0001, 0001
02, 08, 12	1024, 1824, 4224	8576, 6976, 2176	5376, 7376, 3376	9376, 9376, 9376
04, 14, 06, 16	8576, 4976, 6176, 7776	7776, 0576, 2976, 4176	1376, 5376, 7376, 3376	9376, 9376, 9376, 9376

Table 28

BN	BN^5	BN^{25}	BN^{125}	BN^{625}
18, 168, 468, 618	89568, 21568, 85568	OT-3568	213568	813568
32, 132, 232, 882	54432, 42432, 30432	ET-6432	386432	186432
7, 107, 707, 857	16807, 17307, 20307, 96057	00807, 03307, 18307, 169557	20807, 333307, 408307, 539557	420807, 983307, 358307, 889557
43, 143, 493, 893	08443, 08943, 85693, 87693	50443, 52943, 111693, 121693	60443, 72943, 241693, 291693	110443, 672943, 891693, 141693

In these tables we see what happens if we continue with calculating higher powers, especially with the final figures. In the practice of all days nobody will calculate these numbers really, e.g. 6^{500} results in a number of about 390 figures, let alone if we should take bigger numbers!

Look at the rows with 18 and 132, these numbers calculated to the 625th power end on 813568 and 186432, which numbers if added result in 1,000,000. And the difference between 358307 and 420807 is 62,500, a multiple of $15,625 = 5^6$.

There are more interesting things to be seen; in the fifth power the hundreds can be added: 107^5 is $7^5 + 500$; 707^5 ends on 0307, which is 3,500 more than the 6807 of 7^5.

And 493^5 ends on 85,693, 893^5 ends on 87,693 = + 2,000 = + 400 × 5. These numbers taken to the fifth power – we are now in the 25th power – have 111,693 and 121,693, a difference of 10,000 which is 25 × 400. In the same way, 43^{25} and 143^{25} differ by 2,500; 50443 and 52943.

In this table we see that after a lot of powering for the odd numbers except the five the final figures are 0001, and for the even ones the final figures end on 9376.

There is – much – more. E.g. 159^2 ends on (25) 281 and so does 161^3, (4173) 281. Also 241^4 and 239^6 both end on 561, 441^4 and 439^6 both end on 361.

Further: 112^4 and 187^4, differing by 75, in the fourth power end on respectively 936 and 961. 113^4 and 188^4, also differing by 75, in the fourth power end on respectively 361 and 336.

Finally, if we take the exponents in powers of 10, we get as last figures for:

- The even numbers 109376
- The fives 890625
- The sum of them 109376 + 890625 = (1)000001

17 Table powering 10 and higher

BN	BN^{10}	BN^{11}	BN^{12}	BN^{13}	BN^{14}	BN^{15}	BN^{16}	BN^{17}	BN^{18}	BN^{19}	BN^{20}	BN^{21}
2	024	048	096	192	384	768	536	072	144	288	576	152
3	049	147	441	323	969	907	721	163	489	467	401	203
7	249	743	201	407	849	943	601	207	449	143	001	007
8	824	592	736	888	104	832	656	248	984	872	976	808
9	401	609	481	329	969	649	841	569	121	089	801	209
11	601	611	721	931	241	651	161	771	481	291	201	211
12	224	688	256	072	864	368	416	992	904	848	176	112
13	849	037	481	253	289	757	841	933	129	677	801	413
17	449	633	761	937	929	793	481	177	009	153	601	217
18	624	232	176	168	824	432	776	968	424	632	376	768
23	649	927	321	383	809	607	961	103	369	487	201	623
27	649	523	121	267	209	643	361	747	169	563	201	427
33	449	817	961	713	529	457	081	673	209	897	601	833
37	849	413	281	397	689	493	241	917	929	373	801	637
42	824	608	536	512	504	168	056	352	784	928	976	992
43	249	707	401	243	449	307	201	643	649	907	001	043
44	176	744	736	384	896	424	656	864	016	704	976	944
47	049	303	241	327	369	343	121	687	289	583	401	847
48	024	152	296	208	984	232	136	528	344	512	576	648
49	001	049	401	649	801	249	201	849	601	449	001	049
51	001	051	601	651	201	251	801	851	401	451	001	051
52	024	248	896	592	784	768	936	672	944	088	576	952
53	049	597	641	973	569	157	321	013	689	517	401	253
56	176	856	936	416	296	576	256	336	816	696	976	656
57	249	193	001	057	249	193	001	057	249	193	001	057
59	401	659	881	979	761	899	041	419	721	539	801	259
62	224	888	056	472	264	368	816	592	704	648	176	912
67	449	083	561	587	329	043	881	027	809	203	601	267
72	424	528	016	152	944	968	696	112	064	208	776	872
73	649	377	521	033	409	857	561	953	569	537	201	673
77	649	973	921	917	609	893	761	597	969	613	201	477
79	201	879	241	839	281	199	721	959	761	119	401	679
83	449	267	161	363	129	707	681	523	409	947	601	883
84	776	184	456	304	536	024	016	344	896	264	176	784
87	849	863	081	047	089	743	641	767	729	423	801	687
89	601	489	521	369	841	849	561	929	681	609	201	889
91	401	491	681	971	361	851	441	131	921	811	801	891
92	824	808	336	912	904	168	456	952	584	728	976	792
93	249	157	601	893	049	557	801	493	849	957	001	093
97	049	753	041	977	769	593	521	537	089	633	401	897
98	024	352	496	608	584	232	736	128	544	312	576	448

BN means Basic Number

In this table you can recognise the words of Wim Klein; the difficulty of a big root is not in the exponent but in the magnitude of the answer. E.g. the 23th root of a 45 figure number like

685 441 117 453 157 943 684 715 454 813 971 461 595 561 769

seems to be a high grade performance. The experts know it is very simple: there is only one number which fits. In the table you can see that 89^{13} ends on 3 **69** and 369 × 601 (89^{10}), then you have 89^{23}, results in 221 **769**. Even without this table a question like this is not more than an easy picking.

In this table we make an analysis about what happens when powering. This is a very interesting matter, so let's "milk it".

We consider the last two figures, this is normal, sometimes the last three figures. Attention please; not all the two figure numbers are listed!

Then some remarkable things happen: if the tenth power of the number ends on 01, we see that in the following powers the sum of the numbers corresponds with the exponent. E.g. 9^{11} ends on 609, 91^{11} ends on 491, their sum is 100. This happens in the same way with all the numbers of which the basic numbers end on 1 or 9. The tenth power of these numbers always ends on 01. And 01 × any number ends on that number: 01 × 21 ends on 21, 01 × 71 ends on 71.

Then we see that in the eleventh power the sum of two complementing numbers is 100, e.g. 9 and 91; in the thirteenth power the sum of a group is 300, in the same way up to the nineteenth power, where the sum is 900.

There is another group where the basic number ends on 3 or 7. There the tenth power ends on 49. And 49 is a kind of an "inverter", a little bit different from 99. 99 × any number results in final figures equal to 100 minus that number: 99 × 37 ends on 63. And 49 × any numbers results in 50 minus that number. So 49 × 37 ends on 13.

2	-	98	=	024	-	024	=	000
3	-	97	=	049	-	049	=	000
8	-	92	=	824	-	824	=	000
17	-	83	=	449	-	449	=	000
23	-	27	=	649	-	649	=	000
47	-	53	=	949	-	049	=	000
48	-	52	=	024	-	024	=	000
49	-	51	=	001	-	001	=	000

In the table we see that for 7 and 93 the following happens: In the eleventh power the sum of the last figures of these numbers – 743 and 157 – is 900. 49 functions as an inverter; 3 × 9 ends on 7.

The numbers ending in 2 or 8 to the tenth power end on 24. As 24 is also a kind of inverter; 48 × 24 results in 52, so 48 and 52 behave comparably with 7 and 93.

The sum of 2^{11} and 98^{11}, 048 + 352 = 400.

17.1 About the 10th power

In the squares of these groups there are differences of some hundreds; they are all even, based on $(a+b)^2$. An even hundred, multiplied by five, results in a thousand. In the same way, this takes place when a square is calculated to the fifth power. This explains the 000 column.

We will also see that when powering rather often the tens and the hundreds are carried over.

17.2 About the 11th power

Again we consider the last two figures.

2	+	98	=	048	+	352	=	400
3	+	97	=	147	+	753	=	900
8	+	92	=	592	+	808	=	400
17	+	83	=	633	+	267	=	900
23	+	27	=	927	+	523	=	450
47	+	53	=	327	+	973	=	300
48	+	52	=	152	+	248	=	400
49	+	51	=	049	+	051	=	100

The sums of the eleventh powers gives us:

- For the even numbers the sum is 400
- For the odd numbers, the sums result in 900. Except 1 – 99 and 49 – 51; their sum is 100.
- $9 + 11 = 20.$ $9^{11} + 11^{11} = 609 + 611 = 220$
- $49 + 51 = 100.$ $49^{11} + 51^{11} = 49 + 51 = 100$
- $89 + 91 = 180.$ $89^{11} + 91^{11} = 980$

17.3 About the 12th power

2	-	98	=	096	- 496	= 600
3	-	97	=	441	- 041	= 400
8	-	92	=	736	- 336	= 400
17	-	83	=	761	- 161	= 600
23	-	27	=	321	- 121	= 200
47	-	53	=	241	- 641	= 600
48	-	52	=	296	- 896	= 400
49	-	51	=	401	- 601	= 800

The squares of 2, 004 and 98, 9604 differ by 600. In the twelfth power we see the same difference between the hundreds. The other numbers follow this line.

17.4 About the 13th power

2	+	98	=	192	+ 608	= 800
3	+	97	=	323	+ 977	= 300
8	+	92	=	888	+ 912	= 800
17	+	83	=	937	+ 363	= 300
23	+	27	=	383	+ 267	= 650
47	+	53	=	327	+ 973	= 300
48	+	52	=	208	+ 592	= 800
49	+	51	=	649	+ 651	= 300

$2 + 98 = 100$, as the number pairs are even, in the thirteenth power the sum of the hundreds is an even one, so 800. In the same way we see 8 and 92, also 48 and 52.

For the odd number pairs, as $3 + 97 (= 100)$ and $17 + 83 (= 100)$ we see that in the thirteenth power the hundreds are multiplied by (1)3, the sums are 300.

For $23 + 27 = 50$ we see that the sums in the thirteenth power are $13 \times 50 = 650$.

- $7 + 93 = 100$. $7^{13} + 93^{11} = 407 + 893 = 300$.
- $9 + 11 = 20$. $9^{13} + 11^{13} = 329 + 931 = 260$
- $23 + 27 = 50$. $9 + 11 = 20$. $23^{13} + 27^{13} = 383 + 267 = 650$

17.5 About the 14th power

2	-	98	=	384	-	584	=	800
3	-	97	=	969	-	769	=	200
8	-	92	=	104	-	904	=	200
17	-	83	=	929	-	129	=	800
23	-	27	=	809	-	209	=	600
47	-	53	=	369	-	569	=	800
48	-	52	=	984	-	784	=	200
49	-	51	=	801	-	201	=	600

The squares of 2 and 98, 004 & 9604, differ in this way by 400. In the fourteenth power, the difference in hundreds = (2)800 = 7 × 400. Compare the pairs 8 and 92; 48 and 52; 200 is the complement of 800.

The odd number pairs 17 / 83 and 47 / 53 differ by 600 in the squares, and seven times this in the fourteenth power, 200 again; 200 is the complement of 800.

17.6 About the 15th power

2	+	98	=	768	+	232	=	000
3	+	97	=	907	+	593	=	500
8	+	92	=	832	+	168	=	000
17	+	83	=	793	+	707	=	500
23	+	27	=	607	+	643	=	250
47	+	53	=	343	+	157	=	500
48	+	52	=	232	+	768	=	000
49	+	51	=	249	+	251	=	500

We see here a repetition of what we saw before. The sums of the fifteenth power of the even numbers is some thousand, 000.
For the odd numbers is so that 15 × 100 = (1)500, the sums are all 500.

17.7 About the 16th power

2	-	98	=	536	-	736	=	800
3	-	97	=	721	-	521	=	200
8	-	92	=	656	-	456	=	200
17	-	83	=	481	-	681	=	800
23	-	27	=	961	-	361	=	600
47	-	53	=	121	-	321	=	800
48	-	52	=	136	-	936	=	800
49	-	51	=	201	-	801	=	400

Here the differences are practically equal to the difference in the fourteenth power. 48 / 52 differ by 800, the complement of 200. In the same way 49 / 51 differ by 400, the complement of 600.

17.8 About the 17th power

2	+	98	=	072	+	128	= 200
3	+	97	=	163	+	537	= 700
8	+	92	=	248	+	952	= 200
17	+	83	=	177	+	523	= 700
23	+	27	=	103	+	747	= 850
47	+	53	=	687	+	013	= 700
48	+	52	=	528	+	672	= 200
49	+	51	=	849	+	851	= 700

- $2 + 98 = 100$. $2^{17} + 98^{17} = 072 + 128 = 200$
- $3 + 97 = 100$. $3^{17} + 97^{17} = 700$, difference 500
- $9 + 11 = 20$. $9^{17} + 11^{17} = 569 + 771 = 340$
- $17 + 33 = 50$. $17^{17} + 33^{17} = 177 + 673 = 850$

17.9 About the 18th power

2	-	98	=	144	-	544	= 600
3	-	97	=	489	-	089	= 400
8	-	92	=	984	-	584	= 400
17	-	83	=	089	-	489	= 600
23	-	27	=	369	-	169	= 200
47	-	53	=	289	-	689	= 600
48	-	52	=	344	-	944	= 400
49	-	51	=	601	-	401	= 200

17.10 About the 19th power

2	+	98	=	288	+	312	=	600
3	+	97	=	467	+	633	=	100
8	+	92	=	872	+	728	=	600
17	+	83	=	153	+	947	=	100
23	+	27	=	487	+	563	=	050
47	+	53	=	583	+	517	=	100
48	+	52	=	512	+	088	=	600
49	+	51	=	449	+	451	=	900

- $9 + 11 = 20.\quad 9^{19} + 11^{19} = 089 + 291 = 380$
- $89 + 91 = 180.\ 89^{19} + 91^{19} = 609 + 811 = 420\ (19 \times 180 = 3\ 420)$

17.11 About the 20th power

The twentieth power is for all the numbers the "point of return", as for the odd numbers the last two figures are 01, which means that multiplication by the basic numbers will result in the same last two figures, as e.g. $01 \times 13 = 13$.

The even numbers all end on 76, which implies that multiplication by the basic number also results in the same last figures, as e.g. 76×12 results in 12.

For the good order: I did not aim at completeness, there may be a lot of other surprises in the table.

18 Diophantine Equations

In the book "the great mental calculators" we can find a special question on page 248. The question is: "Find three numbers such that their sum is 43 and the sum of their cubes is 17,299. The numbers are 25, 11 and 7.

The way of solving this question is not mentioned. This fact triggered me to make a study of this type of question and after a lot of thinking I found a solution, in fact without any mathematical form, but by sieving out incorrect solutions long enough for a correct one to be found.

This chapter concerns a special type of Diophantine equations. The name comes from the Hellenistic mathematician Diophantus from Alexandria, who lived ± 250 A.D. who in his studies of these equations was one of the first to introduce symbols in algebra.

Definition: a Diophantine equation is an indefinite polynomial equation of which the variables may only be integer numbers. Diophantine problems have fewer equations than unknown variables.

This definition requires some elucidation.

A polynomial is an algebraic expression with several terms. Terms are things we add, so simply said an addition with letters, for example: $2x + 4y + 3z$, or $4x^3 + 6x^2 + 8x + 9$. This makes an equation if we put behind the polynomial an = ('equals') sign and ask for a result. If there is only 1 letter in the equation, then it is solvable. If there are more unknowns then the equation is not solvable that is: the equation is indefinite.
These, so far, are the official mathematic definitions.

In this chapter we will only speak about Diophantine equations of the following shape, e.g. $a + b + c = 23$; $a^3 + b^3 + c^3 = 2135$. In this case the solution is $a = 12$; $b = 7$ and $c = 4$. We will describe in which way the solution for this type of questions can be found in an arithmetic way. So far this type of question has not appeared in tournaments for mental calculation.

The clue after a lot of thinking was, you will not be surprised, modulo calculation; in this case mod 27.

For the good order: For solving questions like this by heart, a perfect knowledge of the cubes of the numbers up to 100 is indispensable. For finding the solutions with bigger numbers I used paper to keep a clear overview, but in no way a calculator machine or any other such aids.

And again, it is difficult to describe the intense joy I felt, after it appeared that my method works. And even if it takes a lot of "sievings" there is the joy of "yes, again the solution is found".

Well, willing friends became instructions: "Please give me ± 20 numbers according to this question $a + b + c = x$ and $a^3 + b^3 + c^3 = z$ and please limit the sum of $a + b + c$ to ± 100.

Later on I was given the instruction for Mathematica as follows:

```
For [i=1, i=<100, i++
a=RandomInteger[{10,99}]
b=RandomInteger[{10,99}]
c=RandomInteger[{10,99}]
x = a + b + c
y = a³ + b³ + c³
Print[{x,y}]]
```

You can read here that the instructions consist of:

- Integer numbers, randomly chosen
- Numbers of two figures, i.e. 10 – 99
- There are three of them, a, b and c, and they are added, but a, b and c themselves are not mentioned!
- a, b and c are calculated to the third power and added: $y = a^3 + b^3 + c^3$;
- The result is printed

This gives one hundred question numbers to be solved. Theoretically the biggest sum is $97^3 + 98^3 + 99^3 = 2824164$, but this number has not yet been presented to me. The biggest until now is 2186116, the sum of $99^3 + 98^3 + 65^3$. For the good order: the numbers do not necessarily to be different, but in all the numerous questions I printed it always occurred that the numbers were different.

Before diving into the hard practice it is useful to mention something about modulo calculation in the third power. If we work with two numbers in the odd powers the following happens; if we have for example $3^3 + 4^3$, we calculate $3 + 4 = 7$. Then we know that the result of the addition will be divisible by 7: $3^3 + 4^3 = 91$, which is divisible by 7. In the same way, $3^5 + 4^5 = 1267$, which is divisible by 7; finally $3^7 + 4^7 = 18571$, which also is divisible by 7.

But this principle does not work if we combine three numbers. $3^3 + 19^3 = 6886$, which is divisible by 22. $4^3 + 19^3 = 6923$, which is divisible by 23. Let's now take $3^3 + 4^3 +$

$19^3 = 6950$, which is not divisible by 26.

So if we take the sum of three integer numbers, we cannot deduce something to find any of the three numbers quickly, we have to do it another way.

However, if we know one of the numbers a, b and c, we can do something. Suppose the question is a + b + c = 26 and the sum of their cubes is 6950. After having subtracted 3^3 and have 6923 we may conclude that one of the factors of 6923 will be 19 + 4 = 23. In the same way if a = 4 and we subtract $6950 - 4^3$ we have 6886 and then conclude that one of the factors will be 22.

Now I am going to reveal my method. I hope you can share with me the deep satisfaction I felt after having seen it all works. Before I had the Mathematica instruction some people were so kind as to make questions for me, because as you know there is one "holy" principle; one should never make his own questions. At a given moment I could not find the answer and spent a lot of paper with calculations. As the questions were made by an acknowledged mathematician I doubted on myself rather than on him. Ultimately I gathered all my courage and asked very prudently "Is it possible that there is a mistake in question 14, as I cannot find the solution". Reaction "Excuse me, I made a mistake in the addition of the cubes, this is the correct number". And then, within a minute finding the solution, yes this gives a very rich satisfaction…..

Ok, we work with modulo 27, according to the table hereunder.
Before starting with mod 27 it is good to study the "conversion table mod 9 / mod 27". Be prepared for surprises – a lot of possibilities will be worked out.

Table 1

X	X³	Mod 27	X	X³	Mod 27	X	X³	Mod 27
1	1	1	10	1000	1	19	6859	1
2	8	8	11	1331	8	20	8000	8
3	27	0	12	1728	0	21	9261	0
4	64	10	13	2197	10	22	10648	10
5	125	17	14	2744	17	23	12167	17
6	216	0	15	3375	0	24	13824	0
7	343	19	16	4096	19	25	15625	19
8	512	26	17	4913	26	26	17576	26
9	729	0	18	5832	0	27	19683	0

The following table presents the complete work out of modulo 27:

Table 2

Mod 9 cubed	Gets mod 27	Mod 9 cubed	Gets mod 27
1	1	5	17
2	8	6	0
3	0	7	19
4	10	8	26

Studying this table we find:

- 9 times we have 0
- The moduli 1, 8, 10, 17, 19 and 26 each appear 3 times
- 1 + 26 = 0; 8 + 19 = 0; 10 + 17 = 0, of course all mod 27
- What remains are the moduli 1, 8, 10, 17, 19 and 26.

But do not forget: Our problems concern the sum of three of these moduli. Based on this 4, 5, 13, 14, 22, 23 as sums are impossible. Anything else is possible. So, we have 21 possibilities, and 6 impossible modulo 27 outcomes. In a table the possibilities are to be seen.

To make the solutions less 'trial and error' a table was made:

In the column "Sum mod 27" the possibilities are mentioned for the question numbers of the shape $a^3 + b^3 + c^3$ and their mod 27 values. The four columns beside, mod 27, list the possibilities of combinations to reach the sum. E.g. If the question number is 0 mod 27 this can be obtained by the combinations as mentioned. The column "Common" indicates if there is a common mod 27 in the sum of moduli. If so, this is the best way to start with the common mod 27.

Table 3 Modular Values for Diophantine Equations

Mod 27	Combinations producing the total							#	Common
0	0, 0, 0	0, 1, 26	0, 8, 19	0, 10, 17				4	0
1	0, 0, 1	1, 1, 26	1, 8, 19	1, 10, 17	8, 10, 10	10, 19, 26	17, 19, 19	7	N/A
2	0, 1, 1	0, 10, 19						2	0
3	1, 1, 1	1, 10, 19	10, 10, 10	19, 19, 19				4	N/A
6	8, 8, 17	8, 26, 26	17, 17, 26					3	N/A
7	0, 8, 26	0, 17, 17						2	0
8	0, 0, 8	1, 8, 26	1, 17 17	8, 8, 19	8, 10, 17	10, 26, 26	17, 19, 26	7	N/A
9	0, 1, 8	0, 10, 26	0, 17, 19					3	0
10	0, 0, 10	1, 1, 8	1, 10, 26	1, 17, 19	8, 10, 19	10, 10, 17	19, 19, 26	7	N/A
11	0, 1, 10	0, 19, 19						2	0
12	1, 1, 10	1, 19, 19	10, 10, 19					3	N/A
15	8, 8, 26	8, 17, 17	17, 26, 26					3	N/A
16	0, 8, 8	0, 17, 26						2	0
17	0, 0, 17	1, 8, 8	1, 17, 26	8, 10, 26	8, 17, 19	10, 17, 17	19, 26, 26	7	N/A
18	0, 1, 17	0, 8, 10	0, 19, 26					3	0
19	0, 0, 19	1, 1, 17	1, 8, 10	1, 19, 26	8, 19, 19	10, 10, 26	10, 17, 19	7	N/A
20	0, 1, 19	0, 10, 10						2	0
21	1, 1, 19	1, 10, 10	10, 19, 19					3	N/A
24	8, 8, 8	8, 17, 26	17, 17, 17	26, 26, 26				4	N/A
25	0, 8, 17	0, 26, 26						2	0
26	0, 0, 26	1, 8, 17	1, 26, 26	8, 8, 10	8, 19, 26	10, 17, 26	17, 17, 19	7	N/A

N/A means not applicable

18.1 How to calculate the mod 27 of a number

For numbers below 1,000: subtract so many times 27 until a number remains smaller than 27. Over 1,000: $27 \times 37 = 999 + 1 = 1,000$. Therefore for every thousand subtract so many times 27 until you have a remainder, smaller than 27. So 13,000 is 13 mod 27. And 13,085 is 13 mod 27 + 85 = 4 mod 27, so the complete number is 17 mod 27.

For 13,085,219? $13 \times 1 = 13$ mod 27, as $10^3 \times 10^3$ equals 1×1 mod 27 this means that 085, in fact thousands, is 4 mod 27 and $219 = 8 \times 27 +$ remander 3, so 3 mod 27, the complete number is $13 + 4 + 3 = 20$ mod 27.

As the practice is the best instructor we start by working out step by step some of the numbers offered to me.

During the 2015 "Rekenwonderweekend" one theme of discussion was: "Can there be found a formula with which we can find without trying one of the three numbers we are so desperately looking for?", and the answer is NO.

Apart from the indispensable modulo calculation we can see rapidly two things:

- The rounded cube root of the question number determines the maximum of 'a'.
- Divide the QN by 3 and calculate the rounded cube root. This is the minimum of 'a', and between those two numbers lies one of the numbers we look for.

18.2 $a^3 + b^3 + c^3 = 684$

Problem: $a^3 + b^3 + c^3 = 684$ and $a + b + c = 18$

The rounded cube root of 684 is 8+, so a (max) = 8. A (min): 684 ÷ 3 = 228, the cube root of it is 6+. So one of our candidates lies between 6 and 8.

As $a^3 + b^3 + c^3 = 684$, which is 9 (mod 27), the first search is to see how can we obtain 9 (mod 27) by adding 3 cubes according to table 3? These are the possibilities:

- 0 + 1 + 8
- 0 + 10 + 26
- 0 + 17 + 19

We see 0 is the common number, we now know that the cube of one of the numbers a, b, and c must be divisible by 27.

As a = 6, the cube is 216. Next we subtract 684 – 216 = 468 ≡ 9 (mod 27). These are the possibilities:

- $6^3 + 6^3 = 216 + 216 = 432$, this is too low, moreover it is conflicting with the idea of three different numbers. Also it is 0 (mod 27)
- $7^3 + 5^3 = 468 \equiv 9$ (mod 27). Here we have the solution

So $684 = 5^3 + 6^3 + 7^3$.

This is in principle the way of working. There is no formula, it is all a matter of sieving out the possibilities until the moment you find the solution. You will see: The bigger the numbers, the more interesting and challenging it gets!!!

18.3 $a^3 + b^3 + c^3 = 2403$

Problem: $a^3 + b^3 + c^3 = 2403$ and $a + b + c = 27$

The rounded cube root of 2403 = 13 +, This is the maximum value of a.

First step: $a^3 + b^3 + c^3 = 0$ (mod 27). A look at table 2 shows that we have these possibilities:

- 0 + 0 + 0
- 0 + 10 + 27
- 0 + 8 + 19
- 0 + 1+ 26

We could be smart and do 27 ÷ 3 = 9 and try a = b = c = 9. But that cock won't fight: 3 × 9^3 = 2187, an important shortfall. One thing is for sure; 0 is the common number in the possibilities, so we first have to look at the numbers whose cubes are 0 (mod 27).

We have 3, 6, 9 and 12. 15 drops out as 15^3 exceeds the question number by far.

Our first try, a = 3.

Then $a^3 + b^3 + c^3 - a^3 = 2403 - 3^3 = 2376$, and $a + b + c - a = 27 - 3 = 24$.

This means that we have to look for two numbers which add up to 24 and of which the sum of their cubes is 2376. We do not need to calculate, even 2×12^3 exceeds 2403 by far.

At this moment it is good to think about the "domination of the biggest number". We take now 13 and 11, their sum is also 24. $13^3 + 11^3 = 2197 + 1331 = 3528$, which exceeds the sum of $12^3 + 12^3$. In the same way $14^3 + 10^3 = 3744$. Further additions are not useful, we can stop here.

Conclusion; a = 3 will not lead to the solution of the question.

Next try, a = 6.

$6^3 = 216 = 0$ (mod 27), of course; now $b^3 + c^3 = a^3 + b^3 + c^3 - a^3$ and $b + c = a + b + c - a = 21$. Can we find two numbers which add up to 21 and for which the sum of their cubes is 2187? The smallest combination is $10^3 + 11^3 = 2331$, so this drops out and by consequence so do all the other possibilities. The "leading" number wins; $12^3 + 9^3 = 2457$, so we are through with 6.

Next one, a = 9.

For a = 9 we have $9^3 = 729$, $b^3 + c^3 = a^3 + b^3 + c^3 - a^3$. $2403 - 729 = 1674$, and $b + c = a + b + c - a = 18$. According to the possibilities we could try $12 + 6$, but already 12^3 exceeds 1674, so it drops out. Then we try $11^2 + 7^3$ and here we are: $1331 + 343 = 1674$ and $11 + 7 = 18$. So now we have the final solution is $2403 = 9^3 + 11^3 + 7^3$ and $9 + 11 + 7 = 27$.

18.4 $a^3 + b^3 + c^3 = 8504$

Problem; $a^3 + b^3 + c^3 = 8504 = 26$ (mod 27) and $a + b + c = 38$.

The rounded cube root of the question number is 20+, so this is the maximum value of a. The minimum value of a = cube root of $(8504 \div 3) = 2834$, thus 14.

As usual, a look at table 3 shows us that the remainder of 26 (mod 27) can be obtained by these additions:

- $0 + 0 + 26$
- $1 + 8 + 17$
- $1 + 26 + 26$
- $8 + 8 + 10$

- $8 + 19 + 26$
- $10 + 17 + 26$
- $17 + 17 + 19$

Based on these possibilities it lies at hand that our first search is for a number which is less than 20 and for which the cube is $= 26 \pmod{27}$. There are $8^3 = 512$ and $512 = 26 \pmod{27}$ and $17^3 = 4913$ and $4913 = 26 \pmod{27}$.

First possibility, $a = 8$ and $8^3 = 512$. Then $b^3 + c^3 = a^3 + b^3 + c^3 - a^3$. $8504 - 512 = 7992$ and $b + c = a + b + c - a = 30$. Next question: Are there two numbers whose sum is 30 and whose cubes when added result in 7992? The lowest possibility is $2 \times 15^3 = 6750$. We can count up with $2 \times 15^3 = 6750$; $16^3 + 14^3 = 6840$; $17^3 + 13^3 = 7110$; $18^3 + 12^3 = 7560$ (each is too few); $19^3 + 11^3 = 8190$, too much.

But we can also use our brains, as follows: The sum of the basic numbers, 30, ends on a zero and so their cubes will do too.

So this solution is a dead end.

Second possibility We go on with $a = 17$ and $17^3 = 4913$;

$b^3 + c^3 = a^3 + b^3 + c^3 - a^3$. $8504 - 4913 = 3591$ and now $b + c = a + b + c - a = 38 - 17 = 21$. Next question: Are there two numbers whose sum is 21 and whose cubes added result in 3591?

It can be one of these combinations: $11 + 10$; $12 + 9$; $13 + 8$; $14 + 7$; $15 + 6$. It is good to think about what will happen if we calculate the sums of these additions.

					Diff − 2331	= × 21	Progr.
10^3	+	11^3	=	2331			
12^3	+	9^3	=	2457	126	6	
13^3	+	8^3	=	2709	378	18	12
14^3	+	7^3	=	3087	756	36	18
15^3	+	6^3	=	3591	1260	60	24

Here we see that the difference between the lowest sum 2331 and the other sums increases in a progressive ("Progr") way. We now can "predict" that the sum of $16^3 + 5^3$ will be $2331 + 90 \times 21 = 2331 + 1890 = 4221$.

Here we can stop as 16^3 exceeds 3591. It is not so very difficult to estimate which is the best "guess": $14^3 + 7^3 = 9 \times 343 = 3087$ and indeed, taking $15^3 + 6^3$ we get the 3591 we are looking for. So the solution is $17^3 + 15^3 + 6^3 = 8504$.

There is also another way of working: we have 3591 and the sum of the basic numbers is 21. At the moment a situation like this appears, that is to say that the sum of the cubes is divisible by the sum of their basic numbers, you can do the following: $3591 \div 21 =$

171. Then $21^2 - 171 = 270$ and $270 \div 3 = 90$.

Now find two numbers which sum is 21 and which product is 90. These will be 6 and 15.

Final solution: the basic numbers are 6, 15 and 17.

So it is crucial that the sum of the two remaining cubes is divisible by the sum of their basic numbers.

Summary:

1. $b^3 + c^3$ must be divisible by $(b + c)$ and define the answer. ($3591 \div 21 = 171$)
2. Square $(b+ c) = 21^2 = 441$
3. Subtract from this square the result of step 1 ($441 - 171 = 270$)
4. Divide (step 3 – step 2) by 3. The result must be divisible by 3. $270 \div 3 = 90$
5. Find the two numbers which are the factors of step 4. $90 = 15 + 6 = 21$.

18.5 $a^3 + b^3 + c^3 = 140939$

Problem: $a^3 + b^3 + c^3 = 140939$ and $a + b + c = 83$

We have $a^3 + b^3 + c^3 = 140939 = 26 \pmod{27}$ The maximum possible number of one of the cubes is a^3, b^3 and c^3 is 52, as $52^3 = 140608$ and 53^3 is already larger than $a^3 + b^3 + c$, as it is 148877.

From table 3 we see that we should start with looking for a number which cube gives us a remainder 26 (mod 27) and smaller than 52.

We make a list of candidate numbers, which cubes are 26 (mod 27). They are: 44, 35, 26, 17 and 8.

First sieve: suppose $a = 44$, so $44^3 = 85184$. $140939 - 44^3$, $85184 = 55755$, and $b + c = a + b + c - a$. $83 - 44 = 39$.

But 55755 is not divisible by 39. This is easy to be seen if you realise that $755 - 55 = 700$ which is evidently not divisible by 13.

Now we try $a = 35$. Then we have $b^3 + c^3 = a^3 + b^3 + c^3 - a^3$; $140939 - 42875 = 98064$ and $b + c = a + b + c - a = 48$.

b cannot be larger than 46, as $46^3 = 97336$. and If $b = 46$, $c = 2$, so $b^3 + c^3 = 97336 + 8 = 97344$. Too small again, therefore 35 drops out as well.

Third sieve: $a = 26$, and $26^3 = 17576$.

Then we have $b^3 + c^3 = a^3 + b^3 + c^3 - a^3$, $140939 - 17576 = 123363$. Now and $b + c = a + b + c - a = 83 - 26 = 57$. B cannot be greater than 49, therefore the maximum of $b^3 + c^3 = 49^3 + 8^3 = 118161$, which is too small. Therefore 26 drops out.

Fourth sieve a = 17, and 17^3 = 4913.

Then we have 140939 – 4913 = 136026. And b + c = a + b + c – a now are 83 – 17 = 66. B cannot be greater than 51, as 51^3 = 132651. If b = 51 then c has to be 15. Now 51^3 + 15^3 = 132651 + 3375 = 136026, and here is our solution.

Or: 136026 ÷ 66 = 2061.
66^2 = 4356 – 2061 = 2295 ÷ 3 = 765.
Find the two numbers whose sum is 66 and whose product is 765.
66 ÷ 2 = 33 and 33^2 = 1089 – 765 = 324.
√324 = 18
Then b = 33 + 18 = 51 and c = 33 – 18 = 15

Final answer 140939 = 17^3 + 51^3 + 15^3.

18.6 To think about

The lowest possible sum of b + c = 66 is 33^3 + 33^3 = 71874.

The table hereunder illustrates the increase of the cubes with b + c = 66.
Basis: (33 + 33) × 3 = 198.

b + c								Increase
33	+	33	=	35937	+	35937	=	71874	= ×198
34	+	32	=	39304	+	32768	=	72072	1
35	+	31	=	42875	+	29791	=	72666	4
36	+	30	=	46656	+	27000	=	73656	9
37	+	29	=	50653	+	24389	=	75042	16
48	+	18	=	110592	+	5832	=	116424	225
49	+	17	=	117649	+	4913	=	122562	256
50	+	16	=	125000	+	4096	=	129096	289
51	+	15	=	132651	+	3375	=	136026	324

In the right hand column we see the increase of the sum of $b^3 + c^3$. The numbers 3, 12 et cetera indicate the difference between 71874 and the following numbers, divided by 66 (the sum of b + c). It will strike you that all the numbers represent X times a square, respectively 1^2, 2^2, 3^2 × 198, et cetera. This column obeys the formula for the difference between the sum of two cubed numbers with a constant sum. From 33 to 50 the difference is 17, and if b = 50 and c = 16, the difference between their cubes is 3 × 17^2 = 867 times 66 = 57222.

In fact it is not necessary to make a walk along all the possibilities. We do not yet know whether a = 17 is indeed the real choice, but it can be. We then calculate b^3+ c^3 =

136026 and go further.

At the moment we are aware that the sum of $33^3 + 33^3$ is far below 136026, we do not need to examine all the numbers following on from 33. Then it is more practical to take a cubed number a bit lower than 136026. Take for example $50^3 = 125000$, add $16^3 = 4096$, then we have 129096. We now know we are "warm". After that the next step $51^3 + 15^3$ is a full hit.

This way of working is identical in all cases where the sum of b + c represents an even number.

18.7 $a^3 + b^3 + c^3 = 359982$

Problem; $a^3 + b^3 + c^3 = 359982$ and $a + b + c = 138$.

We have $a^3 + b^3 + c^3 = 359982 = 18 \pmod{27}$.

The maximum possible number of one of the cubes a^3, b^3 and c^3 is 71 as $71^3 = 357911$. According to table 2 the best approach for this question is to take a number which is divisible by 3, so that $a^3 = 0 \pmod{27}$.

We make a provisional list of some numbers < 71, and then get, counting down, respectively 69, 66, 63, 60, 57, 54, 51, 48. If this list appears to be too short, later on we continue.

Now we start.

a = 69^3 = 328509; 359982 − 328509 = 31.473. and b + c = 69. Here we can stop immediately, as the lowest sum of two cubes, of which the basic number sum is 69, is $35^3 + 34^3$: 42875 + 39304 = 82179. This is far over the mark; full stop.

a = 66. We get $a^3 = 66^3$ and $b^3 + c^3 = a^3 + b^3 + c^3 − a^3 = 359982 − 287496 = 72486$. Now b + c = 72. The lowest possible sum of two cubes whose sum is 72 is $36^3 + 36^3 = 93312$. Again, here we can stop immediately. You'll see; if you look over the edge of your cupboard, a lot of unnecessary efforts can be avoided!

a = 63. We get $a^3 = 250047$. $b^3 + c^3 = a^3 + b^3 + c^3 − a^3 = 359982 − 250047 = 109935$. And b + c = 75. Lowest possible sum of two cubes where the original sum is 75 is $38^3 + 37^3 = 105525$.

In the case of b + c representing an odd number, it works according to the table hereunder. The numbers are cubed.

38	+	37	=	54872	+	50653	=	105525	diff = × 450
39	+	36	=	59319	+	46656	=	105975	1
40	+	35	=	64000	+	42875	=	106875	3
41	+	34	=	68921	+	39304	=	108225	6
42	+	33	=	74088	+	35937	=	110025	10

We take the smallest combination 38 + 37 = 75 and multiply this by 6 and get 450.

The difference is + 2, + 3, + 4, + 5 et cetera. This knowledge will later on be very useful!

A row like this is called "triangle numbers", for which the formula is $\frac{1}{2}n \times (n+1)$. So the fifteenth number in such a row is $\frac{1}{2} \times 15 \times 16 = 120$.

In the meantime we know that a = 63 is not the number we want and continue with a = 60.

a = 60, so a^3 = 216000. So $b^3 + c^3$ = 359982 − 216000 = 143982 and b + c = 78. The lowest possible sum of two cubes whose sum is 78 is $39^3 + 39^3$ = 118638. Max b = 52, as 52^3 = 140608. If b = 51, c = 27 and $b^3 + c^3$ is over the mark. $51^3 + 27^3$ = 132651 + 19683 = 152334, too much. One step down then; $50^3 + 28^3$. As this ends on 00 + 52 = 52 this combination drops off. Also 143982 is not divisible by 78.

Conclusion: a = 60 is not a good choice either.

Another step down, a = 57.

We get a = 57 , so a^3 = 185193, $b^3 + c^3 = a^3 + b^3 + c^3 − a^3$ = 359982 − 185193 = 174789. Now b + c = 138 − 57 = 81. We calculate $54^3 + 27^3$ = 19683 × 9 = 177147. This is too much. A number lower gives $53^3 + 28^3$. Here a rough estimation will do: 53^3 = ± 149000 and 28^3 = ± 22000, this is too low. Besides: 53^3 ends on 877 and 28^3 ends on 952, the sum is 829. We can finish our job with a = 57. Also 174789 is not divisible by 81.

And a = 54? We would get $a^3 = 54^3$ = 157464. $b^3 + c^3 = a^3 + b^3 + c^3 − a^3$ = 359982 − 157464 (54^3) = 202518. If a = 54, then b + c = 84 and $b^3 + c^3$ = 202518. Again an estimation: $56^3 + 28^3$ = 9 × 21952 (28^3) = 197568. One number higher: 57^3 = 185193 + 27^3 = 19683 = 204876. This is over the mark, 54 drops out. Here it is very simple to see that 202518 is not divisible by 84: 18 has 2^1 and 84 has 2^2.

a = 51. We would get a^3 = 132651. $b^3 + c^3 = a^3 + b^3 + c^3 − a^3$ = 359982 − 132651 (51^3) = 227331, and b + c = 87. We estimate $58^3 + 29^3$, which agrees with 9 × 24389 = 219501. We are too low. One step higher, this is more than only a try; 59^3 ends on 79; 28^3 ends on 52, therefore $59^3 + 28^3$ ends in 31. That sounds good!!

And 227331 ÷ 87 = 2613. Then 87^2 = 7569 − 2613 = 4956 ÷ 3 = 1652.

So can we factorise 1652 in such a way that the sum of its factors is 87?

- $87 \div 2 = 43.5^2$ and $43.5^2 = 1892.25$
- $1892.25 - 1652 = 240.25$
- $\sqrt{240.25} = 15.5$
- $a = 43.5 + 15.5 = 59$
- $b = 43.5 - 15.5 = 28$
- $59 \times 28 = 1652$

And indeed, this is a full hit as $59^3 + 28^3 = 227331$. Now we have found the final solution: $a + b + c = 138$ and $51^3 + 28^3 + 59^3 = 359982$.

It all seems to be a plunge in murky water. Not completely; a lot of useless work can be spared if one starts with a well-considered choice. For this table 2 is very helpful.

After having found the solutions to over a hundred of this type of questions, I thought: "Well, it's all clear to me, what can happen?" It is sometimes rather elaborate. E.g. the number 1191445 required eighteen sievings (!!), a lot of work, very challenging, very instructive, but it was all according to the standard path. And it is very instructive, because you learn to filter out the impossible candidates. And then came, purely by coincidence, the number 436510, which appeared to be a very unmanageable number.

18.8 $a^3 + b^3 + c^3 = 436510$, the stumble stone

Problem: $a^3 + b^3 + c^3 = 436510$ and $a + b + c = 154$.

We have $a^3 + b^3 + c^3 = 436510 = 1 \pmod{27}$ and $a + b + c = 154$.
The maximum possible number of one of the cubes a^3, b^3 and c^3 is 75, as $75^3 = 421875$. Following table 2 the attack on the number started with cubed numbers 1 (mod 27). These are the cubes of 73, 64, 55, 46, 37, 28, 19, 10, 1.

The lowest possible number is the cube root of $(436510 \div 3) = 145503+$, so this is 52. Now we know that the numbers 37, 28, 19, 10 and 1 can be ignored.

We start:

$a = 73$. We would get $a^3 = 389017$. $b^3 + c^3 = a^3 + b^3 + c^3 - a^3 = 436510 - 389017(a^3) = 47493$. Then $b + c = 154 - 73 = 81$. The lowest possible sum for $b^3 + c^3 = 41^3 + 40^3 = 132921$, which is far over the mark.

$a = 64$. We would get $a^3 = 262144$. $b^3 + c^3 = a^3 + b^3 + c^3 - a^3 = 436510 - 262144 \,(64^3) = 174366$. $b + c = 90$, the minimum of $b^3 + c^3$ is $2 \times 45^3 = 182250$. Again, over the mark.

We now conclude that one of the numbers, be it a, be it b or be it c, lies between 55 and 46, but is not necessarily one of these two numbers. We start with 55 and then count down, and will stop as soon as we see the number leads to a dead end.

$a = 55$, $a^3 = 166375$. $b^3 + c^3 = a^3 + b^3 + c^3 - a^3 = 436510 - 166375 = 270135$. $b + c = 99$. Minimum sum of $b^3 + c^3 = 50^3 + 49^3 = 125000 + 117649 = 242649$. If $b^3 + c^3 = 60^3 + 39^3 = 275319$; too much. Alternative: $59^3 + 40^3 = 269379$. Too few. So $a = 55$ is not the right approach.

$a = 54$, where $a^3 = 157464$. $b^3 + c^3 = a^3 + b^3 + c^3 - a^3 = 436510 - 157464 = 279046$. $b + c = 100$. Minimum sum of $b^3 + c^3 = 50^3 + 50^3 = 250000$. The approach $b^3 + c^3 = 60^3 + 40^3 = 280000$ is too big, we do not need to calculate. The difference between $280000 - 250000$ is $30000 = 100 \times 300$, the minimum difference between $b^3 + c^3 = 50^3 + 50^3 = 250,000$ and $b^3 + c^3 = 51^3 + 49^3 = 250,300$. The increase of the combinations $b^3 + c^3$ in this case is $x^2 \times 300$. Stepping from 50 to 60 we need 10 numbers, then the increase of $b^3 + c^3$ if $b + c = 100 = 300$ per number.

After this the conclusion can only be 54 also drops out.

Well, next approach: $a = 53$ and a^3 is 148877. What we get now is $b^3 + c^3 = a^3 + b^3 + c^3 - a^3 = 436510 - 148877 = 287633$.

If $a = 53$, then $b + c = 101$. 287633 is not divisible by 101. If if a six figure number is divisible by 101 then it is mandatory that the sum of the first two figures plus the last two figures minus the middle two figures is 0 or 101.

Example: 167963. We take $16 + 63 = 79 - 79 = 0$; $167963 \div 101 = 1663$. In the same way: 287633. Then $28 + 33 = 61 - 76 = -15$, so not divisible by 101.

We now take $b = 60$ and $60^3 = 216000$ and c^3 (41^3) $= 68921$, and finally $b^3 + c^3 = 284921$. Next approach $b = 61$ and $c = 40$, $= b^3 + c^3 = 226981 + 64000 = 290981$.

Here we conclude that $a = 53$ does not result in a correct solution. On to the next number: $a = 52$.

$a = 52$, $52^3 = 140608$. $b^3 + c^3 = a^3 + b^3 + c^3 - a^3 = 436510 - 140608 = 295902$. $b + c = 102$, the minimum sum of $b^3 + c^3 = 51^3 + 51^3 = 265302$. The difference with the combination $52^3 + 50^3 = 140608 + 125000 = 265608$ is 306. We do not need to make all the steps, but make an approach: $295902 - 265302 = 30600$.

Well, that's a surprise: $30600 \div 306 =$ exactly 100 times. As 100 is the square of ten, we now have found that the correct combination for $b^3 + c^3$ is $61^3 + 41^3$: $226981 + 68921 = 295902$.

Another very important thing is looking for the final figures; we go back to $b + c = 102$. If we take e.g. $52^3 + 50^3$ the result ends on 608, more precisely 265608. And $53^3 + 49^3$ end on $77 + 49 \equiv 26$, therefore neither of these combinations is what we are looking for. This is a matter of esteeming and thinking: For obtaining 295902 we can esteem: $60^3 + 40^3 = 280.000$ so we immediately see we have to take bigger numbers, and $61^3 + 41^3$ ends on 902. As there is no combination possible, we calculate this exactly and see it was the correct choice.

We can also do this:

- 295902 ÷ 102 = 2901.
- 102^2 = 10404 – 2901 = 7503
- 7503 ÷ 3 = 2501
- 102 ÷ 2 = 51
- 51^2 = 2601
- 2601 – 2501 = 100
- $\sqrt{100}$ = 10
- a = 51 + 10 = 61
- b = 51 – 10 = 41

The final solution for this question is a = 52, b = 61 and c = 41.

On the internet we can almost daily read something about a world record. This is in fact a convicting proof of ones skills. If one should ask me if this kind of questions lends itself for a world record attempt, I can only say: NO.
Why is that? Well, you see it here above: for finding the numbers we want, there is no straight ahead method. It is a matter of sieving and thinking, and let's not forget: a profound knowledge of the cubes up to hundred. Besides it is a matter of combination of the last figures of a cube to see if the estimation is a good one or if we have to continue.
The deep joy is to be found in the satisfaction of having found the answer on a kind of question for which until now no form exists for finding the answer in one straight ahead move.
And there is more: the well-known English mental calculation prodigy George Lane once wrote that all the kinds of arithmetic operations in one way or another are interlinked. I can only confirm. Also this kind of questions, although they may seem to be absurd, really enrich the feeling with numbers, it is an exercise for training in the knowledge of the cubes, addition, subtraction, division and logical thinking.
And the most: the satisfaction of finding the solution!!

18.9 With the help of modulo 11

During the MCWC (Mental Calculation World Cup) 2016 for adults the honour was granted to me to give an explanation about the diophantine equations and my solutions with the help of modulo 27. This struck the audience. In this select company there are also some mathematicians. Amongst them Dr. Andy Robertshaw and Andreas Berger. They asked me two questions:

- Are you sure that always one solution is possible?
- Did you try some time to find the solution with modulo 11?

My answer on both questions was no. And they started to work. No they did not calculate, they wrote a computer program and this produced a considerable list with two or even more solutions. And with this I started with modulo 11.

This is a complicated affair, but it can be done. It is also very interesting, but a help table is indispensable. Finding the possibilities which the table offers is a separate problem, which requires more time than finding the solution itself. The table is made with the program Mathematica, on instructions of the well known mathematician Dr. B. de Weger. Here it is:

0, {0, 0, 0}, {0, 1, 10}, {0, 2, 9}, {0, 3, 8}, {0, 4, 7}, {0, 5, 6}, {1, 1, 4}, {1, 2, 7}, {1, 3, 3}, {1, 5, 8}, {1, 6, 9}, {2, 2, 8}, {2, 3, 4}, {2, 5, 10}, {2, 6, 6}, {3, 5, 7}, {3, 6, 10}, {3, 9, 9}, {4, 4, 5}, {4, 6, 8}, {4, 9, 10}, {5, 5, 9}, {6, 7, 7}, {7, 8,9}, {7, 10, 10}, {8, 8, 10},

1, {0, 0, 1}, {0, 2, 5}, {0, 3, 6}, {0, 4, 9}, {0, 7, 10}, {0, 8, 8}, {1, 1, 10}, {1, 2, 9}, {1, 3, 8}, {1, 4, 7}, {1, 5, 6}, {2, 2, 6}, {2, 3, 10}, {2, 4, 8}, {2, 7, 7}, {3, 3, 7}, {3, 4, 4}, {3, 5, 9}, {4, 5, 10}, {4, 6, 6}, {5, 5, 5}, {5, 7, 8}, {6, 7, 9}, {6, 8, 10}, {8, 9, 9}, {9, 10, 10},

2, {0, 0, 7}, {0, 1, 1}, {0, 2, 3}, {0, 4, 5}, {0, 6, 8}, {0, 9, 10}, {1, 2, 5}, {1, 3, 6}, {1, 4, 9}, {1, 7, 10}, {1, 8, 8}, {2, 2, 2}, {2, 4, 6}, {2, 7, 9}, {2, 8, 10}, {3, 3, 9}, {3, 4, 10}, {3, 5, 5}, {3, 7, 8}, {4, 4, 8}, {4, 7, 7}, {5, 6, 7}, {5, 8, 9}, {5, 10, 10}, {6, 6, 10}, {6, 9, 9},

3, {0, 0, 9}, {0, 1, 7}, {0, 2, 8}, {0, 3, 4}, {0, 5, 10}, {0, 6, 6}, {1, 1, 1}, {1, 2, 3}, {1, 4, 5}, {1, 6, 8}, {1, 9, 10}, {2, 2, 4}, {2, 5, 7}, {2, 6, 10}, {2, 9, 9}, {3, 3, 5}, {3, 6, 7}, {3, 8, 9}, {3, 10, 10}, {4, 4, 6}, {4, 7, 9}, {4, 8, 10}, {5, 5, 8}, {5, 6, 9}, {7, 7, 10}, {7, 8, 8},

4, {0, 0, 5}, {0, 1, 9}, {0, 2, 6}, {0, 3, 10}, {0, 4, 8}, {0, 7, 7}, {1, 1, 7}, {1, 2, 8}, {1, 3, 4}, {1, 5, 10}, {1, 6, 6}, {2, 2, 10}, {2, 3, 7}, {2, 4, 4}, {2, 5, 9}, {3, 3, 3}, {3, 5, 8}, {3, 6, 9}, {4, 5, 7}, {4, 6, 10}, {4, 9, 9}, {5, 5, 6}, {6, 7, 8}, {7, 9, 10}, {8, 8, 9}, {8, 10, 10},

5, {0, 0, 3}, {0, 1, 5}, {0, 2, 2}, {0, 4, 6}, {0, 7, 9}, {0, 8, 10}, {1, 1, 9}, {1, 2, 6}, {1, 3, 10}, {1, 4, 8}, {1, 7, 7}, {2, 3, 9}, {2, 4, 10}, {2, 5, 5}, {2, 7, 8}, {3, 3, 8}, {3, 4, 7}, {3, 5, 6}, {4, 4, 4}, {4, 5, 9}, {5, 7, 10}, {5, 8, 8}, {6, 6, 7}, {6, 8, 9}, {6, 10, 10}, {9, 9, 10}},

6, {0, 0, 8}, {0, 1, 3}, {0, 2, 4}, {0, 5, 7}, {0, 6, 10}, {0, 9, 9}, {1, 1, 5}, {1, 2, 2}, {1, 4, 6}, {1, 7, 9}, {1, 8, 10}, {2, 3, 5}, {2, 6, 7}, {2, 8, 9}, {2, 10, 10}, {3, 3, 6}, {3, 4, 9}, {3, 7, 10}, {3, 8, 8}, {4, 4, 10}, {4, 5, 5}, {4, 7, 8}, {5, 6, 8}, {5, 9, 10}, {6, 6, 9}, {7, 7, 7},

7, {0, 0, 6}, {0, 1, 8}, {0, 2, 10}, {0, 3, 7}, {0, 4, 4}, {0, 5, 9}, {1, 1, 3}, {1, 2, 4}, {1, 5, 7}, {1, 6, 10}, {1, 9, 9}, {2, 2, 7}, {2, 3, 3}, {2, 5, 8}, {2, 6, 9}, {3, 4, 5}, {3, 6, 8}, {3, 9, 10}, {4, 6, 7}, {4, 8, 9}, {4, 10, 10}, {5, 5, 10}, {5, 6, 6}, {7, 7, 9}, {7, 8, 10}, {8, 8, 8},

8, {0, 0, 2}, {0, 1, 6}, {0, 3, 9}, {0, 4, 10}, {0, 5, 5}, {0, 7, 8}, {1, 1, 8}, {1, 2, 10}, {1, 3, 7}, {1, 4, 4}, {1, 5, 9}, {2, 2, 9}, {2, 3, 8}, {2, 4, 7}, {2, 5, 6}, {3, 3, 4}, {3, 5, 10}, {3, 6, 6}, {4, 5, 8}, {4, 6, 9}, {5, 7, 7}, {6, 7, 10}, {6, 8, 8}, {7, 9, 9}, {8, 9, 10}, {10, 10, 10}},

9, {0, 0, 4}, {0, 1, 2}, {0, 3, 5}, {0, 6, 7}, {0, 8, 9}, {0, 10, 10}, {1, 1, 6}, {1, 3, 9}, {1, 4, 10}, {1, 5, 5}, {1, 7, 8}, {2, 2, 5}, {2, 3, 6}, {2, 4, 9}, {2, 7, 10}, {2, 8, 8}, {3, 3, 10}, {3, 4, 8}, {3, 7, 7}, {4, 4, 7}, {4, 5, 6}, {5, 7, 9}, {5, 8, 10}, {6, 6, 8}, {6, 9, 10}, {9, 9, 9}},

10, {0, 0, 10}, {0, 1, 4}, {0, 2, 7}, {0, 3, 3}, {0, 5, 8}, {0, 6, 9}, {1, 1, 2}, {1, 3, 5}, {1, 6, 7}, {1, 8, 9}, {1, 10, 10}, {2, 2, 3}, {2, 4, 5}, {2, 6, 8}, {2, 9, 10}, {3, 4, 6}, {3, 7, 9}, {3, 8, 10}, {4, 4, 9}, {4, 7, 10}, {4, 8, 8}, {5, 5, 7}, {5, 6, 10}, {5, 9, 9}, {6, 6, 6}, {7, 7, 8}.

And so it works: the big printed number is the modulo 11 of the sum of the three cubed numbers, the so called "big sum". Behind this you'll find "triplets"; without exception for each big sum there are 26 of them. Between accolades are the moduli 11 of the basic numbers, which are not calculated to the third power. If you do so and add the results, you'll get the mod. 11 of the big sum.

Worked out: the mod 11 of the big sum is 10. The mod 11 of the small sum (SS), the three basic numbers, is 5. Then you find 2, 6 and 8: 2 + 6 + 8 = 16 = 5 mod 11. And add the three numbers cubed: 8 + 216 + 512 = 736 = 10 mod 11.

Now you are sufficiently prepared for a real question. We take 226962 as the big sum, the cube root of it is 60 + something. This number is 10 mod 11. The small sum is a + b + c = 108 = 9 mod 11. The table gives as possibilities a, b and c are 0, 2 and 7, together 9 mod 11. The same is valid for the possibilities 1, 3 and 5, and 4, 8 and 8. Here the small sum is also 9 mod 11.

Big sum (B.S)	BN = 0 mod 11	BN^3	BS − BN^3	SS − BN	Int. Div.
226962	55	166375	60587	53	No
	44	85184	141778	64	No
	33	35937	191025	75	2547
	22	10648	212314	86	No
	11	1331	225631	97	No

From the table we take the integer division and do small sum 108 − 33 = 75. Then 75^2 − 2547 = 3078. Next 3078 ÷ 3 = 1026. Finally we look for two numbers of which the sum is 75 and the product is 1026.

The factorization of 1026 = 2 × 3^3 × 19. After a bit of puzzling you'll find that 18 and 57 are the only factors which meet the requirements of the sum 75 and the product 1026.

To check if it all works perfectly we now take a basic number 2 mod 11.

Big sum (B.S)	BN = 0 mod 11	BN³	BS – BN³	SS – BN	Int. Div.
226962	57	185193	41769	51	819
	46	97336	129626	62	No
	35	42875	184087	73	No
	24	13824	213138	84	No
	13	2197	224765	95	No
	2	8	226954	106	No

We see the integer division $41769 \div 51 = 819$ and do $51^2 - 819 = 1782$ and $1782 \div 3 = 594$. Now we look for two numbers of which the product is 594 and of which the sum is 51. The factorization of $594 = 2 \times 3^3 \times 11$ and after some puzzling we find easily that the numbers required are 18 and 33.

We examine the remaining possibility of 7 mod 11.

Big sum (B.S)	BN = 0 mod 11	BN³	BS – BN³	SS – BN	Int. Div.
226962	51	132651	94311	57	No
	40	64000	162962	68	No
	29	24389	202573	79	No
	18	5832	221130	90	2457
	7	343	226619	101	No

With the number 18 we find the integer division with $221130 \div 90 = 2457$. Now $90^2 = 8100$ and $8100 - 2457 = 5643$ and $5643 \div 3 = 1881$. We look for two numbers which' sum is 90 and which' product is 1881 and find easily the numbers 33 and 57.

The table offers also the possibility that the basic numbers are 1, 3 and 5, sum 9 mod 11. We take, randomly, 3 mod 11.

Big sum (B.S)	BN = 0 mod 11	BN³	BS – BN³	SS – BN	Int. Div.
226962	58	195112	31850	50	637
	47	103823	123139	61	No
	36	46656	180306	72	No
	25	15625	211337	83	No
	14	2744	224218	94	No
	3	27	226935	105	No

With $108 - 58 = 50$ we find the integer division has the result 637. Next, $2500 - 637 = 1863$ and $1863 \div 3 = 621$. Which numbers have a sum of 50 and the product 621? These numbers are 23 and 27. So 226962 is the sum of the cubes of 58, 27 and 23.

Andy and Andreas, thank you very much for your interesting questions!!!!

19 Mixed operations

In this chapter we'll talk about the combination of powering and multiplication and the interesting result of this. If you would ask me what will be the practical use of this, I can not give you an answer; I really do not know. However, I believe that all kinds of operations with numbers are interlinked in one way or another and that also this chapter will enrich your insight into the magic world of numbers. You'll see that the results are very often comparable with those of addition.

$11^2 \times 2 = 242$, and $50 - 42 = 8$, almost $11 - 2$.
$11^2 \times 3 = 363$, and $11 + 3 =$ almost $13 = 63 - 50$.

11^3	1331	×	997	=	1327007
111^3	1367631	×	897	=	1226765007
211^3	9393931	×	797	=	7486963007
311^3	30080231	×	697	=	20965921007
411^3	69426531	×	597	=	41447639007
511^3	133432831	×	497	=	66316117007
611^3	228099131	×	397	=	90555355007
711^3	359425431	×	297	=	106749353007
811^3	533411731	×	197	=	105082111007
911^3	756058031	×	97	=	73337629007

$11^3 \times 2 = 2662$, almost $11 + 2$, in this case $12 + 50$. $61^3 \times 2 = (4539)\ 62$, almost $61 + 2$. $11^3 \times 3 = 3993$, and $11 - 3 =$ almost 7, $1{,}000 - 7 = 993$. $11^3 \times 997 = (1307)\ 007$, almost $997 + 11$.

What we see here is very interesting: in all the cases the answers end on 007 (a well-known number), still more striking is the jump between the thousands: this is 8. We go from 7007 to 5007, to 3007 et cetera.

11^3	1331	×	797 =	1060807
111^3	1367631	×	697 =	953238807
211^3	9393931	×	597 =	5608176807
311^3	30080231	×	497 =	14949874807
411^3	69426531	×	397 =	27562332807
511^3	133432831	×	297 =	39629550807
611^3	228099131	×	197 =	44935528807
711^3	359425431	×	97 =	34864266807
811^3	533411731	×	9997 =	5332517074807
911^3	756058031	×	9897 =	7482706332807

It is obvious that if we use the "antipodes" of the numbers, e.g. 3 instead of 997, the last figures will be 3993, in the same way the combination of $111^3 \times 103$: which ends on (14086)5993, again a jump of 2000, now as a plus.

In this table we see that the jump of the thousands is 2,000 and that both $11^3 \times 797$ and $511^3 \times 297$ end on 0807, a symmetry effect.

We also see a jump of the thousands, from 0807 to 8807 to 6807, so the jump of the thousands here is 8,000, or – if you want – minus 2,000.

A comparable table can be made with 61. In combination with 03, 53 et cetera it works in an "inverse" way, in combination with 47, 97 et cetera, it works in a "synchronous" way.

61^3 in combination with 03 and 53 gives a kind of an inverse subtraction: for 61 – 3 you think 57, and 000 – 57 = 943.

			ends on
61^3	×	03	943
	×	53	993
	×	153	093
	×	653	593
	×	47	107
	×	97	157
	×	997	057

BN	SQU				Jump 1	Jump 2
1	1	×	2 =	2		
11	121	×	12 =	1452	450	
21	441	×	22 =	9702	250	800
31	961	×	32 =	30752	50	800
41	1681	×	42 =	70602	850	800
51	2601	×	52 =	135252	650	800
61	3721	×	62 =	230702	450	800
71	5041	×	72 =	362952	250	800
81	6561	×	82 =	538002	50	800
91	8281	×	92 =	761852	850	800

61^3 in combination looks like an addition: for 61+ 47 you think 107, and see $61^3 \times 47$ ends on 107.

In these two small tables we'll recognise again some interesting structures, the magic world of numbers is full of them.

The "hares are not running randomly" in numberland.

As many tables can be made as one likes, it is endless. E.g. $41^2 \times 53$ ends on (89)093; if we multiply by 153 we get (257) 193, so the jump per hundred is 100.

We can take $41^3 \times 37 = (2550)\ 077$; if we take $41^3 \times 337$ we get (23226) 377.

The number 41 is also an interesting number to work – or to play – with. Also here powering and addition resemble each other, with only a difference of 1, see the "41" table.

			ends on
41^2	×	03	043
	×	53	093
	×	153	193
	×	653	693
	×	47	007
	×	97	057
	×	997	957

			L3F				L3F
41^2		3 =	043	41^3	×	37 =	077
	×	53 =	093		×	87 =	127
	×	103 =	143		×	137 =	177
	×	153 =	193		×	187 =	227
	×	203 =	243		×	237 =	277
	×	903 =	943		×	937 =	977
	×	953 =	993		×	987 =	027

What we see here with 41^2 is about the inverse of what we wrote about 61^3. The columns L3F means "Last three figures", where we see e.g. $41^2 \times 53$ ends on 093 et cetera; adding 41 + 53 gives 94. And in the case of $41^3 \times 87$ we get as the last three figures 127, whereas adding gives 128.

If we take $41^3 \times 37$ we get 077 et cetera. Another example: 93^5 ends on (6,956,88) 3693. And if we take $93^3 \times 193 \times 493$ we add 5 × 100, and see the result of this multiplication ends on (76,533,76) 4193, so an increase of 500.

$13^3 \times 17 =$	(3) 7349
$113^3 \times 117 =$	(16881) 8949
$213^3 \times 217 =$	(207900) 0549
$313^3 \times 317 =$	(972058) 2149

Again, do not ask me what this table is good for, I really do not know. But we can simply see the structure in it; the hundreds are in this case increasing by 1600, and you can be sure that e.g. $713^3 \times 717$ will end on 8549. Try it! It is my experience that if a given structure returns three times, it will do this forever. So far, despite furious attempts I have not yet found any exception on this rule.

20 Currency conversion

Did you want to buy a book from amazon.com, prices in $? Or something from England where they calculate in £? Or something from Europe, where the Euro is the standard currency? Here is the information about currency.

There are two kinds of calculation:

- The currency conversion rates, generally calculated to 4 decimal places
- The calculation of the amounts, generally to 2 decimal places.

Inevitably in these both cases you have to cope with the rounding rules: if the first digit beyond the required accuracy is 5 or higher, you should round **up** your answer. If it is lower than 5, then you have to round **down**. So you should realise that you have always to find one digit more than the required accuracy.

The division is made according to what is described in the chapter "divisions". There is absolutely no difference with a "common" division; the only difference is in the sign of the currency unit and not in the outcome!
We have £ 1 = € 1.1155.
Work out: $1.00 \div 1.1 = 8$ r 12, answer so far 0.8
$12 \times 10 = 120 - 8 \times 1 = 112 \div 11 = 9$ r 13, answer so far 0.89
$13 \times 10 = 130 - 9 \times 1 - 8 \times 5 = 81 \div 11 = 6$ r 15, answer so far 0.896
$15 \times 10 = 150 - 6 \times 1 - 9 \times 5 - 8 \times 5 = 59 \div 11 = 4$ r 15, answer so far 0.8964
$15 \times 10 = 70 - 4 \times 15 - 6 \times 5 - 9 \times 5 = 71 \div 11 = 5$ r 16, answer so far 0.89645, rounded to 0.8965
So this conversion rate means that for every Euro I receive £ 0.8965..
In the same way we can find that if £ 1 = $ 1.3452, so an American citizen receives for each $1 "only" £ 0.74338, rounded £ 0.7434.

20.1 Calculation of the amount

An American citizen who wants to buy an article priced at £ 12 will have to pay $12 \div 0.7434 = \$ 16.142$, rounded down in this case, because 2<5, to $ 16.14.

20.2 Conversion rates

- £ 1 = € 1.1633
- € 1 = $ 1.3334
- $ 1 = SF 1.0007

These conversion rates are given; we are to make the required multiplications and divisions to find the answers to the questions.

The conversion rate from £ to $ means that we have to multiply 1.3334 × 1.1633, which is in fact a multiplication of 2 five digit numbers with an answer of 1 digit and 8 decimals, the result: 1.55114422. so for £ 1 an Englishman receives $ 1.5511 for every £1.

Depending on your multiplication abilities you can do this: Make the multiplication 1.33 34 × 1.16 33 as follows: 33 × 34 = 11 22. Forget the 22 as this rate of accuracy is not required and carry the 11. After having taken 33 × 133 = 4389 + 11 = 4400, then do 34 ×116 = 3944 and add 4400 + 3944 = 8344. Then 133 × 116 = 15428 + 8344 = 1551144, with the comma we have 1.551144. Our rounded answer is now 1.5511. We should not ignore the 44 completely, because this can later on play a role in another multiplication, but the last 22 can be ignored.

33 × 34				11	22
16 × 34 + 33 × 33			16	33	
1 ×33 + 1 × 34 + 33 × 16		5	95		
1 × 16 + 1× 33		49			
1 × 1	1				
Total	1	55	11	44	(22)

In fact we may also ignore the last 4, as this does not affect the required accuracy. 1.0007 × 1.551144 = 1.5522 (25798) and 1.0007 × 1.55114 = 1.5522. Finding the conversion rate from £ to SF (Swiss Francs) means 1.0007 × 1.55114 = 1.552230, rounded 1.5522).

€ 1 = SF 1.0007 × 1.3334 = SF 1.3343(333), rounded 1€ = SF 1.3343

20.3 Again: calculation of the amount

£ 6.62 in $ = 6.62 × 1.5511 = 10.268 = rounded to $ 10.27. One can also do the following; firstly take 6.62 × 1.55, result 10.2610 and then 6.62 × 0.0011 results in 0.0072 + something, 10.2610 + 0.0072 gives 10.2682, which when rounded results in 10.27.

$ 15.17 in £ = 15.17 ÷ 1.5511 = £ 9.780, rounded £ 9.78

In the same way, £ 7.23 in SF = 7.23 × 1.56 = 11.2788. Then 7.23 × 0.00002 = 0.001446, of which we only need the 0.0014. Then 11.2788 + 0.0014 gives 11.29, and so rounded to SF 11.29.

SF 16.18 in £ = 16.18 ÷ 1.562 = 10.358, rounded 10.36

€ 10.28 in SF = 10.28 × 1.3343. We firstly do 10.28 × 1.33 giving 13.6724. Then 10.28 × 0.0043 giving 0.0044 + something. 13.6724 + 0.0044 = 13.7166 which when rounded = SF 13.72.

It is of paramount importance with the multiplications of yy.yy × 0.00xx to pay attention to the decimals. Do for your own practice some additions with wrongly placed decimals to see how the results differ!!

SF 12.91 in € = 12.91 ÷ 1.3343 = € 9.6820, rounded to € 9.68.

20.4 Calculation with "Dutch" money

This story took place in the restaurant room of the rail station in Groningen. A young couple speaks to me while drinking a cup of coffee. The man is an entrepreneur and starts an interesting arithmetic story. My reaction amazes him "I never experienced so fast calculating as you do". I give a little show. The girl, ± 22 years old, is a fourth year student in management. She travels with 40% student reduction, her ticket cost € 13.40.

"And which is the gross price of your ticket?"
"Sir, I have no idea how to calculate this. I am not good in calculation, fortunately my father is".
"Well, take him with you to your work, otherwise something will go wrong".

21 Calendar calculation

21.1 The day of the week

Because every week unchangeably contains seven days, this will not give us a problem. On a give day we can add or subtract a multitude of seven, the day of the week remains the same. If the 16th of any month falls on a Tuesday, then the 2nd, 9th, 23rd and 30th are all also a Tuesday. We count the weekdays with a seven-remainder as follows:

1	Sunday	5	Thursday
2	Monday	6	Friday
3	Tuesday	7 or 0	Saturday
4	Wednesday		

Attention: there are several manners of counting the weekdays, depending on who is writing about this. So for the Sunday you can find the remainders of 0, 1 or 6. So do not mix up several methods!!

January	1	April	0	July	0	October	1
February	4	May	2	August	3	November	4
March	4	June	5	September	6	December	6

21.2 The months

Where do these offsets come from? Well January $1 = 1 \div 7 =$ remainder 1. Thirty one days later – so called 32 January – in fact February 1, we divide 32 by 7 and have a remainder 4. As February has 28 days – except in the leap years – March has the same offset. After March comes April and we count (after January 1) $+ 31 + 28 + 28 = 90$, + April $1 = 91$, divided by 7 gives a remainder 0; therefore the offset of April is 0. The offsets of all the other months are calculated in the same way.

The offsets are very easy to remember, we divide them in groups of 3 and get $1 - 4 - 4$, the square of twelve, and $0 - 2 - 5 =$ the square of 5. Then $0 - 3 - 6$, the square of six and finally 1 – 4 – 6, for which no memory trick exists.

An example: 19 September 1939.

Date		19
Month, offset		6
Year: 4, times		9
Year: 7, remainder		4
Century	20, no addition	0
Total, and 7 remainder		38, R = 3
Weekday		Tuesday

Calendar calculation is in fact modulo 7 calculation. As there has already been so much literature published about calendar calculation, this will be discussed no further.

In a separate chapter you will find more about calendar calculation, e.g. for calculating a missing month or year.

21.3 The years

For the years, two calculations are required; the numbers have to be divided by:
- 4, to determine the number of leap years, and
- 7, to determine the mod seven remainder after the division.

Example: year 13. Into 13 we can fit 3 × 4, and when 13 is divided by 7 the remainder is 6. Total offset: 3 + 6 = 9 which is 2 mod 7. For 43, we can fit 10 × 4 and the mod 7 remainder is 1. For the year 43 the offset is 10 + 1 = 11 and 11 ÷ 7 results in a remainder 4.

Year	Offset	Year	Offset	Year	Offset	Year	Offset
00	0	28	0	56	0	84	0
01	1	28	1	57	1	85	1
02	2	30	2	58	2	86	2
03	3	31	3	59	3	87	3
04	5	32	5	60	5	88	5
05	6	33	6	61	6	89	6
06	0	34	0	62	0	90	0
07	1	35	1	63	1	91	1
08	3	36	3	64	3	92	3
09	4	37	4	65	4	93	4
10	5	38	5	66	5	94	5
11	6	39	6	67	6	95	6
12	1	40	1	68	1	96	1
13	2	41	2	69	2	97	2
14	3	42	3	70	3	98	3
15	4	43	4	71	4	99	4
16	6	44	6	72	6		
17	0	45	0	73	0		
18	1	46	1	74	1		
19	2	47	2	75	2		
20	4	48	4	76	4		
21	5	49	5	77	5		
22	6	50	6	78	6		
23	0	51	0	79	0		
24	2	52	2	80	2		
25	3	53	3	81	3		
26	4	54	4	82	4		
27	5	55	5	83	5		

Of course you have already have seen quickly that the columns are chosen in a way so that they express clearly a difference or cycle of 28 years. Based on this you can easily find easily smaller numbers from within the same century by subtracting a multiple of 28.

About the structure: We look at the table and look for the years with an offset of 0. We then get 0, 6, 17, 23, 0 and calculate the cadence which gives us 0, 6, 17, 23, 0. And this results in 6, 11, 6 and 5. Adding this gives 6 + 11 + 6 + 5 = 28, and now we see that after 28 years the calendar is the same, which agrees with seven leap years.

We now look at offset 1 and get the years 1, 7, 12, 18, 29 and get the cadence 6, 5, 6, 11.

- Offset 2, the calendar years 2, 13, 19, 24, 30, cadence 11, 6, 5, 6
- Offset 3, the calendar years 3, 8, 14, 25, 31, cadence 5, 6, 11, 6
- Offset 4, calendar years 9, 15, 20, 26, 37, cadence 6, 5, 6, 11
- Offset 5, calendar years 4, 10, 21, 27, 32, cadence 6, 11, 6, 5
- Offset 6, calendar years 5, 11, 16, 22, 33, cadence 6, 5, 6, 11. Look at that; we know this one already.

Resumed: for the years 1, 4 and 6 we have the same cadence; about the years we can now make the following table:

Leap year	Cadence
+ 0	6, 11, 6, 5
+ 1	6, 5, 6, 11
+ 2	11, 6, 5, 6
+ 3	5, 6, 11, 6

We take 16 October 1927 and examine in which years after that October 16 falls on the same weekday. We count: 16 = 2 mod 7, October has the offset 1, 27 ÷ 4 = 6 times, and 27 ÷ 7 gives a remainder of 6. 2 + 6 + 6 + 1 = 15, which is 1 mod 7, a Sunday. As the day 16 October remains the same we can limit ourselves to the calculation of the years. 1927 = 6 x 4, the 7-remainder is 6, so the of 1927 = 6 + 6 = 5 (mod 7). We add 5 and get 32 which is 8 × 4 and seven remainder 1, and so the offset is 5. Ok. 1932 + 6 = 1938, giving 9 times 4 and 7 remainder 3, and 9 + 3 = 12 which has the seven remainder 5. From 1938 + 11 we get 1949, which is 12 times 4 and has a seven remainder 0, again the offset is 5. We now have 1927, 1932, 1938 and 1949 and can increase these numbers by 28 each.

Attention please: In a leap year things go differently; we take 16 February 1924. We calculate: 16 + 4 − 1 + 6 + 3 = 28, which is 0 mod 7, so this is a Saturday. To get another Saturday we now must look for 16 + 4 + 1 to get 0 mod 7. This means the year 1929. Which year next? If we add 6 and come to 1935 we are correct, as 35 ÷ 4 goes 8 times and has a 0 remainder, just what we need! Adding 11 years results in 1946 into which fit 11 times 4, plus there is a mod seven remainder 4, and 4 + 4 is 1 mod 7. 1946 + 6 = 1952, again a leap year and moreover 1924 + 28 years. So we have 16 October on a Saturday in 1924, 1929, 1935, 1946, 1952 etc.

21.4 The centuries

In the centuries we calculate as belonging to the same century all the years therein from

00 until 99. Concerning the centuries it is valid that the 00 year is not normally a leap year. Only if the century year is divisible by 400, this year is a leap year; for example 1600 and 2000.

We have to consider carefully that there is a "counting" difference: the years 19xx belong to the 20th century. The transition to a following century works as follows: We go from January 1 in 1900 to January 1 in 2000. Then we have 100 "common" days, as a year has 365 days, i.e. 364 (divisible by 7) + 1. And we have 25 "leap days", the addition gives 100 + 25 = 125 which is 6 mod 7. Therefore we have to add 6 days from the 20th century to the 21st one.

The year 2000 is a leap year, and 2100 is not. From 2000 up to 2100 there are 100 "normal" extra days + 24 "leap days", total 124 days = 5 mod 7. And 6 + 5 = 4 (mod 7), so the offset for the years 21xx is 4 in comparison with the years 19xx.

For the centuries we have this table:

Years	Century	To add	Years	Century	To add
1600-99	17	6	2000-99	21	6
1700-99	18	4	2100-99	22	4
1800-99	19	2	2200-99	23	2
1900-99	20	0	2300-99	24	0

This goes on endlessly and can be summarized in these rules:

- Century divided by 4, remainder 0: add 6
- Century divided by 4, remainder 1: add 4
- Century divided by 4, remainder 2: add 2
- Century divided by 4, remainder 3: add 0

From the above mentioned can be concluded that the calendar is repeated equally every 400 years. The explanation: After 400 years we get 400 extra days, plus 100 leap days, but minus 3, because three "century numbers" are not divisible by 4, in total + 497 days, and 497 ÷ 7 gives the remainder 0. Having examined it all, you can conclude that the calendar is logically composed. In fact it is all modulo seven calculation in optima forma.

It is interesting to trace what happens if there is a transition to the next century, starting with a leap year.

We take February 19 1872. We add: 19 + 4 (the offset for February) -1 (leap year) + 18 (72 ÷ 4) + 2 (remainder of 72 ÷ 7) + 2 (19th century) = 44, which means a Monday. Adding 5 gives 1877, and the addition is 19 + 4 + 19 + 2 = 44; again a Monday. Next: +6 = 1883, the addition is 19 + 4 + 20 + 6 + 2 = 51, ÷ 7 gives a remainder of 2, which is Monday. Then 1883 + 11 is 1894, when the addition is 19 + 4 + 23 + 3 + 2 = 51, which is 2 mod 7, a Monday.

Next: 19 February 1900 is 19 + 4 + 0 = 23, so 2 mod 7, Monday. Now the next one. Will that be 1905, 1906 or 1911? 1905 means + 5 days + 1 (times division by 4) = + 6: no! 1906: This has 6 as seven remainder + 1 (times 4), total 7, so 1906 is the correct year. According to the table we can take + 5 or + 11. + 5 results in 1911 and we have 2 (× 4) + 4 (seven remainder), total 6. No, we want a 0 remainder.

1917 means 4 (times 4) + 3 (seven remainder) and 4 + 3 = 0 (seven remainder), so 1917 is correct.

We summarise the years we found for which February is on a Monday: 1872, 1877, 1883, 1894, 1900, 1906, 1912, 1917, 1923, 1928.

21.5 The questions

Calculate the weekday of:

- 18 November 1931
- 27 February 1948
- 16 January 1883
- 22 August 1717
- 12 May 1639
- In which month(s) of the year 1947 did the 23rd fall on a Friday?
- In which month(s) of the year 1816 did the 25th fall on a Thursday?
- In which month(s) of the year 1937 did the 9th fall on a Tuesday?
- In which years between 1843 and 1861 did September 12 fall on a Wednesday?
- In which years between 1960 and 1980 did July 30 fall on a Saturday?

Finally a very interesting one, which I did unfortunately not think of myself. It comes from Jan van Koningsveld and copied, with his approval, from his book "Become a human calendar in 7 days"

- In which years between 2008 and 2024 did the first of the month fall three times on a Monday?

21.6 The answers

- 18 November 1931: 18 + 4 + 3 = offset 4, a Wednesday
- 27 February 1948: 27 + 4 − 1 + 12 + 6 = 48, offset 6, a Friday
- 16 January 1883: 16 + 1 + 20 + 6 + 2 = 45, offset 3, a Tuesday
- 22 August 1717: 22 + 3 + 4 + 3 + 4 = 36, offset 1, a Sunday
- 12 May 1639: 12 + 2 + 9 + 4 + 6 = 33, offset 5, a Thursday

- In which month(s) of the year 1947 did the 23rd fall on a Friday? Friday has offset 6, 23 has offset 2; 1947 has offset 11 (×4) + 5 (7 remainder) = 16, and so an offset of 2. We look for a month with offset 6 – 4 = 2 and this is only May.
- In which month(s) of the year 1816 did the 25th fall on a Thursday?
- The offset of 25 – ?? – 1816 = 25 + 4 (×4) + 2 (seven remainder of the division 16 ÷ 7) + 2 (19th century) has to be 5. Yet we have already 5, so we need months with an offset 0. These are April and July. But wait! 1816 is a leap year; for the months January and February we have to subtract 1 day. So now January has also an offset 0, and the solution is: January, April and July 1816.
- In which month(s) of the year 1937 did the 9th fall on a Tuesday? 9 has an offset 2; 1937 = 9 (×4) + 2 (7 remainder). 2 + 9 + 2 = offset 6 and Thursday has an offset 3. We miss a further offset of 4, and so we find the months February, March and November.
- In which years between 1843 and 1861 did 12 September fell on a Wednesday? We have 12 September, offset 12 + 6 = 4. And Wednesday already has an offset of 4. But attention please: we are in the 19th century, which means + 2. We now have offset 6 and look for years with offset 5 to get an offset of 4. 1843 has offset 10 (i.e. 10 times 4) + 1 (seven remainder) + 2 (19th century), total 13, thus an offset of 6. 1844 = 11 + 2 = offset 6, so that's a no; 1845 = 11 + 3 = offset 0, no; 1846 = 11 + 4 = offset 1, no; 1847 = 11 + 5 = offset 2, no; 1848 = 12 + 6 = offset 4, no; 1849 = 12 + 0 = offset 5, this is what we want! The cadence table gives us +5; 1860 = 15 + 4, offset 5. The years we looked for are 1849, 1855 and 1860.
- In which years between 1960 – 1980 did July 30 fall on a Saturday? July 30 has offset 2 and Saturday has 7; there is a shortfall of 5. We start with 1960 which has 15 (as 15 times 4) and 4 (seven remainder), so 1960 is what we need. We add 5 and get 1965 and get 16 + 2, so no. 1966 = 16 + 3 = offset 5, this is again correct. And 1977? Well, it has 19 × 4 and a zero 7 remainder, which is also correct. Now the question is answered; if we should add another 5 years we would jump over 1980. The years we want then are 1960, 1966 and 1977.
- In which years between 2008 and 2024 did/will the first of the month fall three times on a Monday? Looking at the table for the offset of the month we find three months with the same offset: February, March and November. But hereby we have to ignore the leap years, because then for January and February it means minus 1 day! We start to count with 1 February 2009 and get: 1 + 4 + 2 + 2+ 6 = 15, offset 1, a Sunday. So we say no. If we take one more year, we have 2010 and get 1 + 4 + 2 + 3 + 6 = 16, offset 2. Just what we need! We take the table of the years and add 5, 6 or 11 years, and the result has to be offset 5.

2015 = 3 (× 4) + 1 (seven remainder), offset 4, no
2016 = 4 (× 4) + 2 (seven remainder), offset 6, no

2021 = 5 (× 4) + 0 (seven remainder), offset 5, it fits!

We go back to the leap years; for them there is a reduction of 1 day and then January has offset 0. And we find this offset too in the months April and July.

We now calculate the weekday of 1 January 2008: 1 + 1 (offset January) − 1 (for a leap year) + 6 (21st century) + 2 (× 4) + 1 (seven remainder) makes 10, offset 3, a Tuesday. But we look for a Monday, so we test the other leap years up to 2024, by adding 5 for each leap year (5 days). This results in:

- 1 January 2012, Sunday
- 1 January 2016, Friday
- 1 January 2020, Wednesday
- 1 January 2024, Monday

So we found another year and the complete solution for this question is the years 2010, 2021, 2024.

22 Four consecutive primes

The effort "four consecutive primes" was an idea of Mr. Robert Fountain, the man who granted me the very nice epithet "William Flash, King of the Primes".

He estimated me as being the only one worldwide who was capable of this. Ok, it was worthwhile to think about and to try, in fact a real challenge!

The performance itself took place during the rekenwonderweekend on May 12 in 2012. The questions were made by means of a random generator, there were series of three question numbers each with 13, 14, 15 or 16 figures.

For this a ready knowledge of all the primes up to 10,000 was indispensable; 1062 numbers. A table with them is given so that you can easily follow what is done when finding the correct answers.

About the half of them, 528, are the sum of two different squares in one way. They are all 1 mod 4, the other ones are 3 mod 4.

The effort concerns numbers from 13 up to 16 digits, which are the product of four consecutive prime numbers of four digits each.

These numbers are not learnt by heart, gradually they got stored in my memory: it will be about 60 years that prime numbers have played a role in my work with numbers.

As there is no competition in this field, it is not a matter of world record, it is a matter of challenge.

22.1 About the 4 figure primes

If you look at the table immediately one thing is striking: the length of the rows differs; it is a bit like a wave of water. Only sometimes are the rows of two neighbour columns equally long: 1900 and 2000, also 2200 and 2300; there are even three equally long columns in 4500, 4600 and 4700; the same with 7700, 7800 and 7900. The maximum number of primes is in the 1400 column (17); the smallest number are in the 5900 and 9500 columns: only 7 primes in each.

In the 5900 and 9500 are also the least 1 mod 4 primes: only 2 in each.

The primes which are the sum of two different squares in one way are printed in bold. Also there are differences, sometimes rather big ones. The biggest difference is 8467 – 8501 = 34; with a hundred differences the only other one that big is 1361 – 1327. Also sometimes there is none: in 1330, 1340 and 1350 there is not a single prime! Within the hundreds there are quite a lot of smallest possible differences, they are not mentioned all: 1019 – 1021; 1031 – 1033; 1049 – 1051; 1229 – 1231; 1301 – 1303; 2267 – 2269 and many more. These numbers are called "twin primes"

In the transition from one hundred to the following one at times the differences are as small as possible: 3299 – 3301; 4799 – 4801; 5099 – 5101; 6299 – 6301; 8999 – 9001.

What also is very interesting is to have a look on the "distances" between the prime numbers. Sometimes there are, within one ten, four prime numbers: 1481, 1483, 1487, 1489; 1871, 1873, 1877 and 1879; 2081, 2083, 2087 and 2089; 5651, 5653, 5657, and 5659 et cetera, to be seen on the table.

In these series I found a link: all the tens are 1 mod 3. Then the ones and sevens are 2 mod 3 and the threes and nines are 1 mod 3.

We'll see later on if the sometimes big "distances" between two prime affect the result of the calculation.

Sometimes there are some consecutive prime numbers which are the sum of two different squares, in the 3400 set there are five in a row! 3413, 3433, 3449, 3457 and 3461. The same happens in same another hundred: 8669, 8677, 8681, 8689, 8693. This is not seen in other hundreds.

Of course a lot of people have tried to find if there is not a certain structure to discover: there is none. Add let's be glad about: if there were a structure, the practical use of the prime numbers would disappear and with this the possibility to use them in the encryption of (amongst others) electronic messages.

The shortest or soonest end is in the 8400 column: this one stops at just 8467.

Let's start with the practice.

The way of working is explained with this thirteen digit number:

$$1\ 6300\ 5476\ 2289$$

Firstly, find the square root of this number, the result is not written on the paper. To give an impression about the accuracy it is important to know how many digits are needed for an acceptable accuracy. We can take the roots of an increasing number of digits to find this out.

Q.N.	SQRT QN	4th root QN
16300	1276.714	1129.9176
163005400	1276.735	1129.9269
163005476	1276.735	1129.9269

In fact it is enough to calculate the fourth root of only five figures of the question number to know in which range the prime numbers wanted are to be found. We know beforehand that the given number is not an exact fourth power.

Here we are in the "about 1130" environment and now are going to multiply with only the last two digits of the primes around 1130. They are 1100 + 09, 17, 23, 29, 51, 53. According to the question number, the result of the multiplication of four consecutive prime numbers has to end on 89. Hereby it will do that we restrict ourselves to the two last figures, as they do not change in the hundreds atmosphere.

Primes 11xx	Multiplication	Last digits
09 × 17 × 23 × 29	53 × 67	51
17 × 23 × 29 × 51	91 × 79	89
23 × 29 × 51 × 53	67 × 03	01

Here you see that the second multiplication gives the desired result, so the final answer for the question number is $1117 \times 1123 \times 1129 \times 1151$.

Another thing which is interesting to know is this: Is there a difference between the arithmetic average of the prime numbers and the rounded fourth root of the question number?

For this, this number was taken:

$$5184\ 4644\ 9221\ 3731$$

Sixteen digits. The fourth root of the complete number is 8485.471441, rounded to 8485. The prime numbers to be found are these ones: 8461, 8467, 8501, 8513. The sum divided by 4 = 8485.5. This means that the rounded fourth root gives a reliable estimation of where the prime numbers can be found.

We take another question number, where the candidates are closer to each other:

$$2\ 7691\ 4560\ 5049$$

Thirteen digits. The fourth root of the complete number is 1289.99031, rounded to 1290. The prime numbers to be found are these ones: 1283, 1289, 1291, 1297. The sum divided by 4 = 1290. This means again that the rounded fourth root gives a reliable estimation of where the prime numbers can be found.

22.2 How to do this?

We discuss the way to work with each kind of question number, i.e. 13, 14, 15 and 16 digits separately, because there is a matter of rounding.

22.3 13 digits

For sufficient accuracy we need only the first five digits of the question number.

Question number	3873711619133	73398238313809
First 5 d	38737	73392
Square root 5 d	196.81	270.91
Fourth root 5 d	1402.9	1645.9
Arithm. Average	1403	1646
Surr. Numbers	13 61, 67, 73, 81, 99, 1409, 23	16 19, 21, 27, 37, 57, 63
Answer	1381 × 1399 × 1409 × 1423	1627 × 1637 × 1657 × 1663

The arithmetic average is the sum of the four primes numbers wanted divided by 4. There is hardly a difference between this number and the rounded fourth root of the question number. This means that the numbers we look for are in the close neighbourhood of the calculated fourth root. This is enlightened by the notion of "surrounding numbers".

For avoiding a wrong answer it is of utmost importance to take the square root of five digits, and for calculation of the fourth root it is also indispensable to take 5 digits.

In a 4th root of a thirteen digit number this has to be subdivided from right to left in groups of four digits. For avoiding a wrong answer it is of utmost importance to take the square root of five digits, and for calculation of the fourth root it is also indispensable to take 5 digits.
So you take $\sqrt{38737}$, which is 196.81 and the square root of 196.81 is 14 + something, so the fourth root of 38737 will be 14 + something.

It is on purpose that there were taken prime numbers with the biggest differences possible; e.g. 1423 − 1381 = 42. Nevertheless the difference between the rounded fourth root and the arithmetic average is negligible.

Once the fourth root has been found it is a matter of "looking around". From 1402.9 (and so 1403) we take some prime numbers below and some ones above and then get e.g. 1403 and below there are 13 61, 67, 73, 81, 99, 1403 and 1423.

Next the multiplication of 61 × 67 × 73 × 81 × 99, looking only at the last two digits, in steps :

61 × 67 = 87 × 81 = 47 × 99 = 43. Not the correct ones.
Then: 67 × 73 = 91 × 81 = 71 × 99 = 29. This try drops off too.
Next: 73 × 81 = 13 × 99 = 87 × 03 = 61. Not the right ones.
Then: 81 × 99 = 19 × 09 = 71 × 23 = 33. Aha, here we are!
The final answer: 3873711619133 = 1381 × 1399 × 1409 × 1423.

Next question number:

$$7\ 3392\ 3831\ 3809$$

The number has 13 digits, the answer has 4, and as $16^4 = 65536$ and $17^4 = 83521$, the answer will be 16 + something.

Square root of 73392 = 270.91, square root of 270.91 = 1645.9, rounded to 1646; this is the area we will work in.

These are the surrounding numbers: 16 19, 21, 27, 37, 57, 63, and we will multiply as mentioned above, until we have the result with the two last digits being 09.

19 × 21 = 99 × 27 = 73 × 37 = 01, not correct.
21 × 27 = 67 × 37 = 79 × 57 = 03, not correct.
27 × 37 = 99 × 57 = 43 × 63 = 09, here we are!
Final answer: 73398238313809 = 1627 × 1637 × 1657 × 1663

22.4 Two 14 digit numbers

The first one:

$$11\ 1650\ 4193\ 8221$$

As $18^4 = 104976$ and $19^4 = 130321$ we know that the first two digits of the answer will be 18.

Question number	1165041938221	22334206337321
First 6 d	111650	223342
Square root 6 d	334.14	472.59
Fourth root 6 d	18.279	21.739
Arithm. Average	1828	2174
Surr. Numbers	18 11, 23, 31, 47, 61, 67	21 53, 61, 79, 22 03, 07, 13

As the complete answers are rather close to the arithmetic average, we now do things a little bit differently: we take this average and go two numbers minus and two numbers plus, and then see what happens.

Working from 1828 the numbers are 1811, 23 (minus) and 31 47 (plus). The multiplications:

$11 \times 23 = 53 \times 31 = 43 \times 47 = 21$. Here we are!!
Final answer: $1165041938221 = 1811 \times 1823 \times 1831 \times 1847$.

Next one:

22 3342 0633 7321

According to the table we will look around the number 2174 and our first manoeuvre is to take in relation to 2174 two prime numbers below and two primes above. We then get 2153, 2161, and 2179, with 2203.

The multiplications: $53 \times 61 = 33 \times 79 = 07 \times 03 = 21$, in one go!!
The complete answer for 22334206337321 is $2153 \times 2161 \times 2179 \times 2203$.

22.5 Two 15 digit numbers

307 3200 3473 2099

Question number	307320034732099	551007403812037
First 7 d	3073200	5510074
Square root 7 d	1753	2347
Fourth root 7 d	41.869	48.449
Arithm. Average	4187	4845
Surr. Numbers	41 57 59, 77, 4201, 11, 17	4813, 17, 31, 61, 71, 77

As above we take two primes lower and two primes higher in relation to 4187 and get this multiplication:

$59 \times 77 = 43 \times 01 = 43 \times 77 = 11$. No, not this one.
Then $77 \times 01 = 77 \times 11 = 47 \times 17 = 99$. Well, this is the right one; the complete answer is $307320034732099 = 4177 \times 4201 \times 4211 \times 4217$.

551 0074 0381 2037

As above we take two primes minus and two primes plus in relation to 4845 and get: 4817, 4831 and 4861, 4871.

The multiplication 17 × 31 = 27 × 61 = 47 × 71 = 37, just what is required.
The final answer is: 551007403812037 = 4817 × 4831 × 4861 × 4871.

22.6 Finally two 16 digit numbers

Here we take the roots of the first 8 digits; four would be too inaccurate.

4053 0930 6423 7781

Question number	4053093064237781	5144270124013789
First 8 d	40530930	51442701
Square root 8 d	6366.39	7172.35
Fourth root 8 d	79.789	84.689
Arithm. Average	7979	8467
Surr. Numbers	79 49, 51, 63, 93, 8009, 11	8443, 47, 61, 67, 8501, 13

As above we take two primes minus and two primes plus in relation to 7979 and get 7963, 7993, 8009, 8011.

The multiplication: 63 × 93 = 59 × 09 = 31 × 11 = 41, no not the right one. We take one prime backward, this is the multiplication: 51 × 63 = 13 × 93 = 09 × 09 = 81 and here is the answer: 4053093064237781 = 7951 × 7963 × 7993 × 8009.

22.7 Finally

5144 2701 2401 3789

For all clarity: it is not guaranteed that the working method "arithmetic average plus and minus two" always and unchangeably in one move leads to the correct answer. But one thing is for sure: we are always very close to it.

Square root of 51442701 = 7172.3567. Square root of 7172 = 84.68, rounded 8467, our starting point.

As above we take two primes minus and two primes plus in relation to 8467 84 43, 47, 61, 67. This is the multiplication, simplified:

43 × 47 ends on 21; 21 × 61 ends on 81 and 81 × 67 ends on 27.
So this is not the correct answer!
Next one: 47 × 61 ends on 67 and 67 × 67 ends on 89 and 89 × 01 ends on 89.
Now we are done!
The answer is: 5144270124013789 = 8447 × 8461 × 8467 × 8501.

Primes up to 5,000

(Table of primes up to 5,000)

22 Four consecutive primes

Primes 5,000 - 10,000

5003	5101	5209	5303	5407	5501	5623	5701	5801	5903	6007	6101	6203	6301	6421	6521	6607	6701	6803	6907	7001	7103	7207	7307	7411
5009	5107	5227	5309	5413	5503	5639	5711	5807	5923	6011	6113	6211	6311	6427	6529	6619	6703	6823	6911	7013	7109	7211	7309	7417
5011	5113	5231	5323	5417	5507	5641	5717	5813	5927	6029	6121	6217	6317	6449	6547	6637	6709	6827	6917	7019	7121	7213	7321	7433
5021	5119	5233	5333	5419	5519	5647	5737	5821	5939	6037	6131	6221	6323	6451	6551	6653	6719	6829	6947	7027	7127	7219	7331	7451
5023	5147	5237	5347	5431	5521	5651	5741	5827	5953	6043	6133	6229	6329	6469	6553	6659	6733	6833	6949	7039	7129	7229	7333	7457
5039	5153	5261	5351	5437	5527	5653	5743	5839	5981	6047	6143	6247	6337	6473	6563	6661	6737	6841	6959	7043	7151	7237	7349	7459
5051	5167	5273	5381	5441	5531	5657	5749	5843	5987	6053	6151	6257	6343	6481	6569	6673	6761	6857	6961	7057	7159	7243	7351	7477
5059	5171	5279	5387	5443	5557	5659	5779	5849		6067	6163	6263	6353	6491	6571	6679	6763	6863	6967	7069	7177	7247	7369	7481
5077	5179	5281	5393	5449	5563	5669	5783	5851		6073	6173	6269	6359		6577	6689	6779	6869	6971	7079	7187	7253	7393	7487
5081	5189	5297	5399	5471	5569	5683	5791	5857		6079	6197	6271	6361		6581	6691	6781	6871	6977		7193	7283		7489
5087	5197			5477	5573	5689		5861		6089	6199	6277	6367		6599		6791	6883	6983			7297		7499
5099				5479	5581	5693		5867		6091		6287	6373				6793	6899	6991					
				5483	5591			5869				6299	6379						6997					
								5879					6389											
								5881					6397											
								5897																
12	11	10	10	13	13	12	10	16	9	12	11	13	15	8	11	10	12	12	13	9	10	11	9	11	281
4	5	7	4	6	6	6	5	9	2	5	6	6	9	5	6	5	7	5	7	4	5	5	6	6	139

7507	7603	7703	7817	7901	8009	8101	8209	8311	8419	8501	8609	8707	8803	8923	9001	9103	9203	9311	9403	9511	9601	9719	9803	9901
7517	7607	7717	7823	7907	8011	8111	8219	8317	8423	8513	8623	8713	8807	8929	9007	9109	9209	9319	9413	9521	9613	9721	9811	9907
7523	7621	7723	7829	7919	8017	8117	8221	8329	8429	8521	8627	8719	8819	8933	9011	9127	9221	9323	9419	9533	9619	9733	9817	9923
7529	7639	7727	7841	7927	8039	8123	8231	8353	8431	8527	8629	8731	8821	8941	9013	9133	9227	9337	9421	9539	9623	9739	9829	9929
7537	7643	7741	7853	7933	8053	8147	8233	8363	8443	8537	8641	8737	8831	8951	9029	9137	9239	9341	9431	9547	9629	9743	9833	9931
7541	7649	7753	7867	7937	8059	8161	8237	8369	8447	8539	8647	8741	8837	8963	9041	9151	9241	9343	9433	9551	9631	9749	9839	9941
7547	7669	7757	7873	7949	8069	8167	8243	8377	8461	8543	8663	8747	8839	8969	9043	9157	9257	9349	9437	9587	9643	9767	9851	9949
7549	7673	7759	7877	7951	8081	8171	8263	8387	8467	8563	8669	8753	8849	8971	9049	9161	9277	9371	9439		9649	9769	9857	9967
7559	7681	7789	7879	7963	8087	8179	8269	8389		8573	8677	8761	8861	8999	9059	9173	9281	9377	9461		9661	9781	9859	9973
7561	7687	7793	7883	7993	8089	8191	8273			8581	8681	8779	8863		9067	9181	9283	9391	9463		9677	9787	9871	
7573	7691				8093		8287			8597	8689	8783	8867		9091	9187	9293	9397	9467		9679	9791	9883	
7577	7699						8291			8599	8693		8887			9199			9473		9689		9887	
7583							8293				8699		8893						9479		9697			
7589							8297												9491					
7591																			9497					
15	12	10	10	11	11	10	14	9	8	12	13	11	13	9	11	12	11	11	15	7	13	11	12	9	279
9	5	6	4	5	7	3	8	6	2	7	8	6	5	4	5	7	5	5	7	2	8	5	4	5	141

total 1062
528

22.7 Finally

259

23 "Der rasche Rückwärtsrechner"

The German TV company ZDF had a program "Superhirn", Super Brain. For this program participants are invited who do something with their brain, regardless what. One could see a lady who, by means of lip reading, could say which aria was being sung from which opera on a silent film. There was also a man who could hear which dance was being danced by the rhythm in the sound of the dancers' feet.

The program is (and the get up and presentation leaves no possibility for an error) a show program. But it is one in which the brain has the leading part. In this activity fits – of course – also mental calculation. Thus was to be seen the Englishman Robert Fountain who calculated the 113th root of a thousand figure number.

Also I was granted the honour to participate. The task concerns two numbers, a and b, each consisting of ten figures. The calculator is presented with their sum, a + b, and their product, a × b.

The participant gets presented a matrix with ninety numbers of ten digits each in increasing magnitude, amongst them the numbers a and b. The task for the participant is to indicate which number is a and which one is b.

It is a nice German alliteration to create for this activity the name "Der Rasche Rückwärts Rechner", three times R. It means "The Fast Backward Calculator". This way of calculation is so interesting that a special chapter is devoted to it. The video about this can be seen on the website willemboumanrekenen.nl in the video gallery.

We start – of course – low profile. And firstly we start with multiplication, because before we calculate backward, we have to calculate forward.

23.1 One common final figure

What is meant here is the combination of two numbers with a common final figure, such as 12 + 22 and 34 + 54.

There is a Dutch childrens' song "Three times three is nine, everyone sings his own song". If we add three + three, so we get six, we know two things:

- There are two numbers a and b, in this case they are equal, and their sum is a + b = 6
- The same numbers are multiplied, a × b = 9

To express this in the mathematical way, we now have two equations and two unknowns:

- (a + b) = 6 and
- (a × b) = 9

As we have two equations and two unknowns we can calculate a and b, the equations are solvable.

We do this according to this method:

1. We divide (a + b) by 2 and then get 6 ÷ 2 = 3
2. We square the result from point 1 and get $3^2 = 9$
3. We subtract (a × b) = 9 from point 2 and then get 9 – 9 = 0
4. We calculate the root of point 3 (0); the result is ± 0
5. We take the number of point 1 (which is 3) and subtract the number of point 4 (i.e. 0) and 3 remains as 3
6. We add the number of point 1 (3) and add the number of point 4 (0), and 3 + 0 = 3

We now have proved that if the sum of two numbers is six, and their product is 9, this only is valid if both numbers are 3.

23.2 A bit of algebra

Now we are doing some mathematics; if we want to obtain the maximum result of the sum of two numbers, this is the way to do it: divide the sum by two and square the result.

We take as an example the three plus three and their product is nine. If we take 4 × 2, their sum is also six, we find as product 4 × 2 = 8, and the shortfall is even more if we take 5 + 1 = 6 and multiply 5 × 1 = 5.

Behind this activity goes what we learnt many years ago at school with algebra under the name "remarkable product"; the formula $(a + b) \times (a - b) = a^2 - b^2$.

Here we can conclude that if we take as the biggest number a and b = 0 then we get the maximum result when squaring. As soon as b is bigger than 0 then is it subtracted from a^2 which by consequence is getting less.

We now can say that if the product of two numbers ends on a 9 and the sum of the numbers ends on 6, then the basic numbers will end on 3.

We make a step forward and start with an odd sum of the two numbers: a + b = 17 and a × b = 52. The solution goes on exactly the same way:

1. (a + b) = 17 ÷ 2 = 8.5
2. 8.5² = 72.25
3. 72.25 − 52 (the product) = 20.25
4. √20.25 = ± 4.5
5. We subtract point 1 minus point 4: 8.5 − 4.5 = 4
6. We add point 1 plus point 4: 8.5 + 4.5 = 13
7. The check: 4 × 13 = 52

Should you shrink from odd numbers because fractions are then inevitable, there is still hope for you. However, it is less nasty than you might think as ?5 × ?5 always results in ?5 and nothing else!

But this is the way to do it:

1. We do not divide by two, but leave the sum unchanged, b = 17
2. We calculate 17² = 289
3. We multiply the product × 4: 4 × 52 = 208
4. We subtract point 2 minus point 3: 289 − 208 = 81
5. We calculate √81 = ± 9
6. We subtract 17 − 9 = 8
7. We add 17 + 9 = 26
8. We divide (attention – do not forget!!) points 6 and 7 by 2 and then get a = 4 and b = 13.

23.3 A bit bigger size

What we revealed here is – of course – also valid with bigger numbers. If two numbers each end on a 3 and their sum is 26 then unchangeably their product ends on **69**. This is not only valid for 13 × 13 = 1 **69**, also 3 × 23 = **69**. But also if the sum of the two numbers is 226, there is the same reasoning, be it that the number of combinations is considerably bigger. Then the smallest product is 3 × 223 = 6 **69**, the biggest product is 113 × 113 = 127 **69**; as you see the last figures are invariably **69**.

And there is the inverse; if you are offered a number ending on 69 and the sum of the factors is xxxxx69 then the possibilities are 3 × 23, 13 × 13, 33 × 93, 43 × 83, 53 × 73, or 63 x 63. In the meantime we are busy with backward calculation.

We continue with forward calculation. We do not only get equal final figures if we add 100 to a given number, adding fifty is already enough! We leave behind the mathematical proof, we restrict ourselves to the practice and take firstly two odd numbers, 11 × 23 = 2 **53.** We add 50 at each number and get 61 and 73. Then 61 × 73 = 44 **53**.

It gets a bit more nasty if we combine an even and an odd number, especially if that even number is only divisible by 2. We now do 14 × 73 = 10 **22** and 64 × 73 = 46 **72**. 4672 − 1022 = 3650**,** the difference is not surprising! But if we take two even numbers it is a different matter: 14 × 24 = 3 **36** and 64 × 74 = 47 **36.**

What also works very well is the use of the complement, in this case up to 100. We have already seen 11 × 23 = 2 **53** and 61 × 73 = 44 **53**. The complements are respectively 89 × 77 = 68 **53** and 39 × 27 = 10 **53**.

Backward calculation: if we look exclusively at the two last figures of the sum we see this: the sum of two numbers is ? 66 and their product is ?? 53. The possibilities are exclusively 89 with 77 and 39 with 27. Other possibilities do not exist concerning this sum and product.

A somewhat bigger combination is a + b = 607 and a × b = 92002.

1. $607^2 = 368449$
2. 4 × 92002 = 368008
3. 368449 − 368008 = 441
4. $\sqrt{441} = \pm 21$
5. We now know a + b = 607 and a − b = 21
6. From here we calculate (607 + 21) ÷ 2 = 314
7. b = (607 − 21) ÷ 2 = 293

What we wrote here above remains valid, regardless the magnitude of the numbers, and we go to the bigger work. Please take a pencil and paper to be able to follow everything precisely.

For completeness: The real matrix in the TV program contained ninety numbers, presented on nine boards with ten numbers each, sorted by magnitude.

We begin: The sum of two numbers is 26 and the table of multiplication gives the results, from bigger to smaller:

13	×	13	=	169	6	×	20	=	120
12	×	14	=	168	5	×	21	=	105
11	×	15	=	165	4	×	22	=	88
10	×	16	=	160	3	×	23	=	69
9	×	17	=	153	2	×	24	=	48
8	×	18	=	144	1	×	25	=	25
7	×	19	=	133					

We see here clearly that with each of the multiplications the sum of the factors is equal, namely 26, and the two last figures of their products are always different.

We now go to calculate backwards with the problem; the sum of two numbers is 26 and their product ends on 53. We know that in principle these numbers can only be 9 and 17, as all the other combinations give different final figures. This is also already known; we may increase each of these numbers by fifty, then the sum is again ? 26 and the product ends on 53 also, e.g. 59 × 67 = 39 **53**.

We hope you can approve that not all the combinations up to 100 are fully elaborated; for example two other numbers, with of course a different sum, may have the same final figures when if multiplied: 13 and 81, their sum is 94 and the multiplication is 10 **53**.

23.4 And now, the real work!

We now take a real training problem and take our time to sift it quietly. About a and b we know this; the sum is 10626122926 and the product is 27681746353669786113. We think: the sum ends on 6 and the product ends on 3, so the last figures can only be 7 and 9. Then ?7 × ?9 are together 26 and their product ends on 13, this is only possible with 47 × 79 = 37 **13** and 29 × 97 = 28 **13**.

So far it's been about the last figures.

Now the first ones. We have a look at the sum, of which we take the first four figures and divide this number by two and get 1062 ÷ 2 = 531. Then 531^2 = 281961 − 276817 (the first six figures of the product) = 5144. $\sqrt{5144}$ = ± 72.

Next we calculate the rounded value of 531 − 72 = 459 and the rounded value of b = 531 + 72 = 603. This is a good approach as 459 × 603 = 276777.
This is the moment to have a look at the matrix and pay attention to "the front side" (on the rounded value) and at the back side (to the last figures).
The table is a part of the matrix, which contains – please remember – ninety numbers.

4390810320	5659498339
4394885771	5822751344
4530570172	5975332381
4561542839	6052571947
4573550979	6151657733
4809001567	6274822082
4927929209	6400600633
4859018696	6467202458
5170231195	6531129933
5217256215	6548106475

We see two numbers which answer to the description: 457 355 0979 and 602 557 1947.

Now there are two data which let us conclude that these numbers indeed are the answers of the question. For the product 79 × 47 gives a product which ends on 13 and if we add 0979 + 1947 we get 2926. The addition 4573 + 6052 Ξ 10625 +, also matches.

Around these numbers there is no other combination to be found which can fulfil these both conditions. Therefore a = 4573550979 and b = 6052571947.

Possibly this questions raises up in you: How difficult are these questions and how much time does it take per question?

The answer: none of these questions is easy and the recorded times vary from a bit less than a minute up to six minutes and seventeen seconds.

23.5 Finally

The mathematically well-grounded reader will recognise the so called abc formula for the solution of quadratic equations. This is valid for finding the first figures of the number to be found.

For finding the last figures there is no formula available, here the only expedient is a profound knowledge of the multiplication of two figure numbers, and the inverse of it. If for eample is given the sum of two numbers ends on 13 and the product ends on 36, the last figures of the numbers can be 04 and 09, but also 24 and 89, 29 and 84, 49 + 64, 69 and 44.

24 Percent calculations

Percent calculation is close to multiplication, nevertheless for elucidation some examples aregiven. Percent is an abbreviation of the Latin pro cent, per hundred.

Example 1
There is a special offer for photo albums, every fourth copy is for half the price. The full price is € 26. There will be ordered seven copies. Which is the average reduction?
We get 3 × 100 + 1 × 50 + 3 × 100 = 650% × € 26 = € 169. The average price is € 24.14.
The average reduction is 50 ÷ 7 = 7.14%.

Example 2
A cargo weighs 746 kilos. The packing is 2%. What is the net weight?
2% of 746 = 14.92 kilo. The net weight then is 746 – 14.92 = 731.08 kilo.

Example 3
The salesman of a TV says he saves only € 68,= which is 4.7 % of the sales price. So the price of the TV was 100 × 68 ÷ 4.7 = € 1,446.80.

Example 4
A dealer buys 2,500 articles of which he sells 1987 pieces with a profit of 35%. In the sales period he sells 473 articles with a profit of 10%, the remaining 40 articles are soldto his own personnel, for the net price. What is his average margin?

1987 × 35 + 473 × 10 = 742.75 ÷ 2500 = 0.2971 × 100 = rounded 30%.

Example 5
In the insurance business they often calculate with pro mille, parts per thousand.
Who insures his furniture for € 230,000 for a premium of 3.07 ‰ pays 3.07 × 230,000 = € 70.61.

Example 6 – Reduction, from what?

In the Netherlands the VAT is 21%, and the companies are obliged to announce their prices for the consumer inclusive VAT.

The net price of an article is € 100, the VAT is 21% so the consumer price is € 121,-.

At times there appear announces in which the company suggests they pay the VAT. In this case the consumer pays € 100,-.

Or it happens that they offer a reduction of 21%, of course calculated from the consumer price. In this case the consumer has to pay € 121 minus 21% = € 25.41 so now he pays € 95.59,-.

So the 21% reduction offer is the most profitable for the consumer.

Example 7

A train ticket with 40% reduction costs € 13.40. Which is the full price?

The best way to do is this: As the reduction is 40%, the price paid represents 60%. Therefore divide 13.40 by 60, so 1.34 by 6 and get 0.2233.

Now multiply 0.2233 × 100 = € 22.33 and round it and get € 22.30.

About the author

Name: Arij Willem Adriaan Pieter Bouman, 19–09–1939.

Education: Primary and secondary school, Institute for the Automotive trade in Driebergen,

Work: 1962–1967 auto engineer, 1967–1984 salesman Michelin truck-tyres, 1985–1995 instructor at the Michelin Centre for Training and Information.

Calculation: all multiplications 2×2 since 1948, 1949 creation of the 11 test, 1952 all squares 0–1000, and square roots of numbers to 10 figures, 1954 cubic roots out of 15 figures, 2006 cubic roots from 18 digits, all integer roots, and formula for jumps at higher powers.

Already as a young child Willem Bouman attracted attention in his surroundings: being 1½ year he knew the alphabet, at 3 he could read the clock. Friends of his parents said "This is not a common child, he is something special". At 9 he knew by heart all the multiplications of 2 × 2 digits. He was taught the nines test, and seeing the restrictions of this he found himself the elevens test. At the secondary school he learnt square roots, which inspired him to examine the structure in the squares, followed by an examination in the cubes. This enabled him to extract integer square roots of 10 digit numbers and integer cube roots of 15 digit numbers. Being no mathematician at all, he completed the secondary school in the department "economy and languages"; he speaks Dutch, English, German and French.

In December 1959 there was a visit to the world famous mental calculator Wim Klein, who was impressed by Bouman's abilities. Despite the advice of Klein to continue in mental calculation, Bouman did not do so; the behaviour of Klein, although very kind, was very nervous and stressed. Bouman did not like to develop himself in this sense.

After service in the army he went to the School for the Automotive Trade and after that he was a motorcar engineer for 6 years; in the evenings he studied to be a teacher in this trade. After that he entered in Michelin as a truck-tyre salesman for the bigger transport companies where his calculation abilities were very helpful, for the calculation e.g. of an axle load, finding the gravity centre of the cargo, the tractive power of a heavy load truck, you name it. He finished his active life as a trainer at the Michelin Centre for Training and information.

After a tip Bouman participated in summer 2006 for the first time and unprepared in the MSO in London, and obtained a 7th ranking. In the same year he participated in Gießen in the Mental Calculation World Cup and obtained a 5th ranking, owing to his profound knowledge of the prime numbers. There he was given the honourable nickname "William Flash, King of the Primes". And there he made acquaintance with a lot of calculation prodigies from all over the world.

IN 2009 during the MSO world championship of mental calculation he obtained a third ranking.

Since 2006 he has been training intensively, which brought him four world records:

- Factorisation, five digit numbers (2010)
- Integer divisions (2011)
- Four consecutive prime numbers from 13 – 16 digits (2012)
- Writing five digit numbers as the sum of four squares (2020)

In October 2014 he was honoured with the 'Prix d'Excellence'.

His major interest is finding new things and structures in the number world.

2 September 1965 he married with Trix van Mourik, the marriage still exists. They have got 3 children, two sons and a girl, and they have 6 grandchildren, 5 girls and 1 boy. None of them inherited his calculation talent. The oldest of them however has excellent school marks; an average of 9 or even more is the standard. The mark for mental calculation of the three oldest granddaughters is 9+.

In January 1995 he could retire. He worked part time in the tyre trade, which stopped in 2003.

Being asked to describe his "fabulous knowledge of mental calculation and methods" he started writing a book about this.

Word of recommendation

According to Stephen Wolfram, the designer of the famous algebra and mathematics software "Wolfram Alpha", mental calculations were not only dull but also irrelevant, and children should not be exposed to it. However much I admire his magnificent software package, nevertheless I disagree with his opinion, which also in our country is blindly adhered to.

"Mental calculation and calculating by hand remain also in this time of iPads and computers of paramount importance. Mathematics would not be there where it is now and computers possibly would not exist if famous mathematicians like Euler, Gauss and Von Neumann would not have commended the profession of mental calculation and calculation by hand and not from childhood had trained intensively with that. Everyone with notion of calculation tuition knows that children like mental calculation and the challenge which entails it.

In the past and in the present there have always been people who have elevated mental calculation to a real art. One of these calculation prodigies is Willem Bouman, the author of this book. He belongs to the best mental; calculators in the world. It is fascinating to take cognizance of the methods which Willem Bouman uses, for example factorisation of big numbers or to calculate the root of it. These and many other baffling calculation methods are presented in this book on an enthusiastic and accessible manor. Cordially recommended for young and old."

Prof. Henk Tijms (President of the Foundation Good Calculation Education).

Thanks to...

To write a book like this without help from outside is an impossible task. Helpers of different shape made it possible to me. The listing is alphabetic:

Eerdewijk, Stephan. Computer expert. Owing to his knowledge many times a file changed into an acceptable form.

Griffioen, Corrinna. Secretary, very skilled in text processing, and moreover, very patient.

Knegt, Arjan de. He lives within walking distance of our house, is always willing to help and many times he just gave that little push which enabled me to continue.

Redactors: Hochstenbach Michiel, Overbeek Harald, Weger, Benne de. Owing to their very valuable remarks, questions and suggestions this book became stature.

Visser, Jasper. Owing to his efforts the edition of this book was possible. He really did a very great job!!

Weger, Benne de. Mathematician at the technical University of Eindhoven for the formulation of the instructions to Mathematica.

Wolfram, Stephen. He invented the academic calculation program "Mathematica". He gave me a license for this program for free, otherwise it would had cost me € 3,267. It is indispensable for the training. If for example I want 100 numbers of 21 figures to calculate the integer root of them, give the correct instruction to Mathematica, shift + enter and in less than no time they are there.

Zwerus, Linda. Photographer. To her we owe the nice photos.

A special word of thanks goes to George Lane, who has two properties which are very helpful to me. He is a

- Native English speaker.
- Mental calculation prodigy, so he understands perfectly my way of thinking.

He was so kind to be active as co-author, to make my book acceptable for the English speaking reader. My English is not – as we call it in Holland – "charcoal English", but not good enough for English standards, and owing to him it can be edited in English.